Lecture Notes in Artificial Intelligence   13114

Subseries of Lecture Notes in Computer Science

More information about this subseries at http://www.springer.com/series/1244

Vincent Lemaire · Simon Malinowski ·
Anthony Bagnall · Thomas Guyet ·
Romain Tavenard · Georgiana Ifrim (Eds.)

Advanced Analytics and Learning on Temporal Data

6th ECML PKDD Workshop, AALTD 2021
Bilbao, Spain, September 13, 2021
Revised Selected Papers

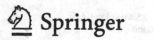 Springer

Editors
Vincent Lemaire 🆔
Orange Labs
Lannion, France

Simon Malinowski 🆔
University of Rennes
Rennes, France

Anthony Bagnall 🆔
University of East Anglia
Norwich, UK

Thomas Guyet 🆔
Inria Grenoble - Rhône-Alpes
Villeurbanne, France

Romain Tavenard 🆔
University of Rennes
Rennes, France

Georgiana Ifrim 🆔
University College Dublin
Dublin, Ireland

ISSN 0302-9743   ISSN 1611-3349 (electronic)
Lecture Notes in Artificial Intelligence
ISBN 978-3-030-91444-8   ISBN 978-3-030-91445-5 (eBook)
https://doi.org/10.1007/978-3-030-91445-5

LNCS Sublibrary: SL7 – Artificial Intelligence

This Springer imprint is published by the registered company Springer Nature Switzerland AG
The registered company address is: Gewerbestrasse 11, 6330 Cham, Switzerland

Preface

The European Conference on Machine Learning and Principles and Practice of Knowledge Discovery in Databases (ECML-PKDD) is the premier European machine learning and data mining conference and builds upon over 19 years of successful events and conferences held across Europe. This year, ECML-PKDD 2021 was planned to take place in Bilbao, Spain, during September 13–17, 2021, but due to the COVID-19 pandemic it was held as a fully virtual event. The main conference was complemented by a workshop program, where each workshop was dedicated to specialized topics, cross-cutting issues, and upcoming research trends. This standalone LNAI volume includes the selected papers of the 6th International Workshop on Advanced Analytics and Learning on Temporal Data (AALTD) held at ECML-PKDD 2021.

Temporal data are frequently encountered in a wide range of domains such as bio-informatics, medicine, finance, and engineering, among many others. They are naturally present in emerging applications such as motion analysis, energy efficient buildings, smart cities, social media, or sensor networks. Contrary to static data, temporal data are of complex nature, they are generally noisy and of high dimensionality, they may be non-stationary (i.e., first order statistics vary with time) and irregular (i.e., involving several time granularities), and they may have several invariant domain-dependent factors such as time delay, translation, scale, or tendency effects. These temporal peculiarities limit the majority of standard statistical models and machine learning approaches, that mainly assume i.i.d data, homoscedasticity, normality of residuals, etc. To tackle such challenging temporal data we require new advanced approaches at the intersection of statistics, time series analysis, signal processing, and machine learning. Defining new approaches that transcend boundaries between several domains to extract valuable information from temporal data is undeniably an important topic and it has been the subject of active research in the last decade.

The aim of the workshop series on AALTD[1] was to bring together researchers and experts in machine learning, data mining, pattern analysis, and statistics to share their challenging issues and advances in temporal data analysis. Analysis and learning from temporal data covers a wide scope of tasks including learning metrics, learning representations, unsupervised feature extraction, clustering, and classification.

For this sixth edition, the proposed workshop received papers that cover one or several of the following topics:

- Temporal Data Clustering
- Classification of Univariate and Multivariate Time Series
- Multivariate Time Series Co-clustering
- Efficient Event Detection
- Modeling Temporal Dependencies
- Advanced Forecasting and Prediction Models

---

[1] https://project.inria.fr/aaltd21/.

- Cluster-based Forecasting
- Explanation Methods for Time Series Classification
- Multimodal Meta-learning for Time Series Regression
- Multivariate Time Series Anomaly Detection

AALTD 2021 was structured as a full-day workshop. We encouraged submissions of regular papers that were up to 16 pages of previously unpublished work. All submitted papers were peer reviewed (double-blind) by two or three reviewers from the Program Committee, and selected on the basis of these reviews. AALTD 2021 received 21 submissions, among which 12 papers were accepted for inclusion in the proceedings. The papers with the highest review rating were selected for oral presentation (seven papers), and the others were given the opportunity to present a poster through a spotlight session and a discussion session (five papers). The workshop had an invited talk on "Deep Generative Models for Missing Data in Temporal Sequences"[2] given by Rose Yu of the UC San Diego department of Computer Science and Engineering, USA[3].

We thank all organizers, reviewers, and authors for the time and effort invested to make this workshop a success. We would also like to express our gratitude to the members of the Program Committee, the Organizing Committee of ECML-PKDD 2021, and the technical staff who helped us to make the virtual AALTD 2021 a successful workshop. Sincere thanks are due to Springer for their help in publishing the proceedings. Lastly, we thank all participants and speakers at AALTD 2021 for their contributions. Their collective support has made the workshop a really interesting and successful event, even under the challenging circumstances of a continuing global pandemic.

November 2021

Vincent Lemaire
Simon Malinowski
Anthony Bagnall
Thomas Guyet
Romain Tavenard
Georgiana Ifrim

---

[2] https://project.inria.fr/aaltd21/invited-speakers/.
[3] https://roseyu.com.

Organization

Program Committee Chairs

Anthony Bagnall	University of East Anglia, UK
Thomas Guyet	Institut Agro, IRISA, France
Georgiana Ifrim	University College Dublin, Ireland
Vincent Lemaire	Orange Labs, France
Simon Malinowski	Université de Rennes, Inria, CNRS, IRISA, France
Romain Tavenard	Université de Rennes 2, COSTEL, France

Program Committee

Amaia Abanda	Basque Center for Applied Mathematics, Spain
Mustafa Baydoğan	Boğaziçi University, Turkey
Alexis Bondu	Orange Labs, France
Paul Honeine	Université de Rouen, France
Antoine Cornuejol	Agro Paris Tech, France
Padraig Cunningham	University College Dublin, Ireland
Dominique Gay	Université de La Réunion, France
David Guijo-Rubio	Universidad de Córdoba, Spain
Iulia Ilie	Siemens, Germany
James Large	University of East Anglia, UK
Brian Mac Namee	University College Dublin, Ireland
Andrei Marinescu	Eaton, Ireland
François Painblanc	Université de Rennes 2, France
Charlotte Pelletier	Université de Bretagne-Sud, IRISA, France
Patrick Schäfer	Humboldt Universität zu Berlin, Germany
Pavel Senin	Los Alamos National Laboratory, USA
Diego Silva	Universidade Federal de Sao Carlos, Brazil
Chang Wei	Monash University, Australia

Contents

Oral Presentation

Ranking by Aggregating Referees: Evaluating the Informativeness of Explanation Methods for Time Series Classification

Surabhi Agarwal, Trang Thu Nguyen, Thach Le Nguyen, and Georgiana Ifrim[✉]

School of Computer Science, University College Dublin, Dublin, Ireland
{surabhi.agarwal,thu.nguyen}@ucdconnect.ie
{thach.lenguyen,georgiana.ifrim}@ucd.ie

Abstract. In this work, we focus on quantitatively evaluating and ranking explanation methods for time series classification based on their informativeness. Time series classification has many applications and evaluating which parts of the time series are most informative for a classifier decision is important. For example, to decide between Arabica and Robusta coffee leaves, we can use an explanation method to highlight the time series parts which differentiate these leaves. Although many explanation methods have been proposed for images and time series data, it is still unclear how to objectively evaluate them. Here, we evaluate two model-specific explanation approaches - ResNet-CAM and MrSEQL-SM, and two model-agnostic approaches, LIME combined with classifiers MrSEQL and ROCKET. We generate saliency-based explanations for each classifier on three time series classification datasets from the UCR benchmark. Importance weights for all points in the timeseries are extracted based on each explanation method, in order to perturb specific parts of the time series and assess the impact on the classification accuracy of referee classifiers. We propose a new ranking-based methodology to compare multiple explanation methods on the basis of their informativeness, by using explanation-based perturbation and aggregating the explanation rank over the referee classifiers. This enables us to compare explanation methods within a single dataset and also across multiple datasets. We provide an in-depth analysis of the results attained, also including runtime analysis for each method. Our results indicate model-specific approaches MrSEQL-SM and ResNet-CAM are much faster than model-agnostic approaches MrSEQL-LIME and ROCKET-LIME and that MrSEQL-SM yields the highest informativeness rank among the explanation methods compared.

Keywords: Time series classification · Explanation methods

© Springer Nature Switzerland AG 2021
V. Lemaire et al. (Eds.): AALTD 2021, LNAI 13114, pp. 3–20, 2021.
https://doi.org/10.1007/978-3-030-91445-5_1

1   Introduction

In recent years Machine Learning (ML) systems have become highly impactful in our everyday life. These methods are growing in terms of their complexity, performance as well as their impact. With the rise in the complexity of ML models, it is also becoming more important to understand their decision-making process which is connected to their *interpretability* [19]. Interpretability is the degree to which a human can understand the cause of a decision [10]. The higher the interpretability of a machine learning model, the easier it is for someone to understand why certain decisions or predictions are made. Understanding the reasons behind these predictions is also important in assessing trust if actions are to be made based on the predictions of the model. Such an understanding gives insights into the model, which can be further used to transform an unstable or inaccurate model or prediction into a stable and trustworthy model [19]. If one can ensure that the ML model can explain decisions and have high interpretability, then the models can be evaluated using some traits such as fairness, privacy, reliability, causality, and trust [7]. The existing approaches can be categorized as techniques that are intrinsic or post-hoc and whether they are global or local [8,16]. A time series is an ordered sequence of numeric values and time series classification (TSC) helps us with predicting a class label for time series. Explainable AI and evaluating the interpretability of TSC methods, help the user understand exactly which part of the time series data resulted in the prediction. This explanation can be visualized as a saliency map by highlighting the parts of the time series which are informative for the classification decision. There are several empirical surveys in recent TSC literature [2,3] and methods which help in designing intrinsic as well as post-hoc explainable models [1,18,19]. However, there is still a strong need to *objectively evaluate and compare* such methods and attain useful explanations. In this work, we evaluate recent explanation methods and propose strategies to provide a *quantitative evaluation using informativeness*. Figure 1 shows the saliency maps produced by four explanation methods: MrSEQL-SM, ResNet-CAM, MrSEQL-LIME and ROCKET-LIME. We can see that the four explanation methods do not agree on which are the important parts of the time series. We aim to evaluate explanation methods based on their informativeness through an explanation-driven perturbation. We focus on methods that produce explanations in the form of saliency maps. In our experiments, we consider two model-specific explanation methods - ResNet-CAM [26] and MrSEQL-SM [11], and two model-independent methods - LIME [19] combined with MrSEQL and ROCKET [4]. The main contributions of this work include:

- A review of the state-of-the-art approaches for explanation of TSC including model-specific explanation methods such as ResNet-CAM and MrSEQL-SM and model-agnostic explanation methods such as LIME and Shapley.
- A new ranking-based methodology to compare multiple explanation methods on the basis of their informativeness, by using explanation-based perturbation and aggregating the explanation rank over a set of referee classifiers.

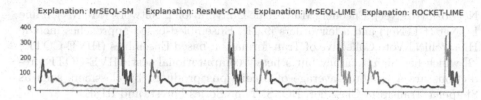

Fig. 1. Saliency map explanations for a motion time series from the dataset CMJ. The most informative parts are highlighted in deep red and the non-informative parts in deep blue. (Color figure online)

- Generation of explanations using LIME for the recent efficient time series classifier ROCKET.
- An empirical analysis of the runtime and an in-depth quantitative evaluation and discussion of the results of four TSC explanation methods ranked over three UCR datasets.

2 Related Work

We first discuss the recent literature on TSC algorithms followed by explanation methods for TSC as well as some of the approaches used to evaluate these explanations.

2.1 Time Series Classification

Time series are commonly used for representing data such as stock prices, weather readings, and biological observations. Time Series Classification (TSC) is a technique used to predict class labels for a given time series [12] and has many applications. In the survey [3] TSC methods have been categorized into five categories including distance-based, interval-based, dictionary-based, ensemble-based, and Deep Learning (DL) based classifiers. The traditional **distance-based classification** technique uses distance measures to determine the class membership. The 1-Nearest-Neighbour algorithm is used as a baseline classifier to classify univariate time series using Euclidean distance and Dynamic Time Warping (DTW) as well as multivariate time series using Frobenius distance [20]. **Interval-based classifiers** select one or more intervals of the series to generate results. An example of interval-based classifiers includes Time Series Forest Classifier (TSF) which adapts the random forest classifier to series data [5]. **Dictionary-based classifiers** form counts of string patterns and then build classifiers based on the resulting features [3]. With the introduction of Bag of SFA symbols (BOSS) [21], Word Extraction for Time Series Classification (WEASEL) [23], SAX-VSM [11] and MrSEQL [11], dictionary-based classifiers have seen major advancements.

Other important classes of TSC algorithms are **DL-based classifiers** and **Ensemble-based classifiers**. DL-based approaches include the use of Multi-Layer Perceptron (MLP), Fully Convolutional Neural Network (FCN), Residual

Network (ResNet), Encoder, Multi-scale Convolutional Network (MCNN), Time Le-Net (t-LeNet) and a few others [9,11]. Ensembled-based approaches include Hierarchical Vote Collective of Transformation-based Ensembles (HIVE-COTE) [2] which has high accuracy but a heavy computational cost. HIVE-COTE predictions are a weighted average of predictions produced by classifiers such as Shapelet Transform Classifier, BOSS, Time Series Forest, and RISE.

2.2   Explanation Methods for Time Series Classification

The goal of an explanation is to relate the feature values of an instance to its model prediction in a way that is understandable to humans [16]. One such tool to represent these explanations is a saliency map.

Saliency Maps. A saliency map is a heatmap that highlights parts of an input that most influenced the output classification [17]. Saliency maps can be used in TSC to highlight the parts of the time series that are important. They are often generated by matching a time series with a vector of weights (w) using a colour map. This vector of weights contains a corresponding weight value for each data point in the time series. The process of generating saliency maps in TSC and producing the vector of weights for the mapping is called the TSC explanation method, and the saliency map produced is known as the TSC explanation [17]. Figure 2 shows a visual representation of how a shape can be converted into a time series using an example of a *Verbena urticifolia* leaf as shown in [25]. The authors of [11] use this representation to classify the Coffee dataset and to produce explanations for the classifier decision as shown in Fig. 3. The highlighted regions of the image correspond to the caffeine and chlorogenic acid components of the coffee blends Arabica and Robusta. An explanation approach has three important aspects as highlighted in [16]:

- Intrinsic or post-hoc: Intrinsic models are those which are considered interpretable due to their simplicity, such as linear models or decision trees. Post-hoc models are black-boxes and special methods need to be developed to obtain explanations.
- Model-specific or model-agnostic: Model-specific approaches are specific to a single model or a group of models. These rely on the working capabilities of the particular model to provide explanations. On the other hand, model-agnostic approaches can be utilized for any ML model regardless of the complexity of the model.
- Local or global scope: The scope of the model can be either local or global depending on whether the method explains an individual prediction or the entire model.

Recent work [17] has shown some contribution towards a quantitative approach for evaluating explanation methods for TSC, such as CAM, MrSEQL-SM and LIME. That methodology proposed an explanation-based perturbation to compute informativeness, but did not provide a way to directly compare and rank explanation methods within and across datasets. In this work, we focus on two model-specific approaches - ResNet-CAM and MrSEQL-SM and two model-agnostic approaches - MrSEQL-LIME and ROCKET-LIME - in order to quantitatively evaluate and rank these methods based on their informativeness.

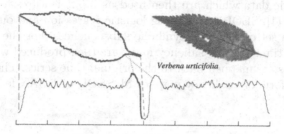

Fig. 2. An example of how a shape can be converted into a *time series* representation (reprinted from [25]).

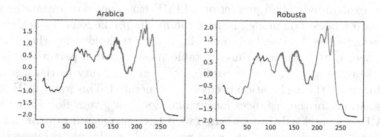

Fig. 3. Saliency mapping generated using the MrSEQL classifier proposed in [11] on the Coffee dataset (reprinted from [11]).

2.3   Model-Specific Approaches

ResNet-CAM. Class Activation Map (CAM) is a model-specific explanation method that helps in explaining the output predictions of a neural network. In previous work [26], CAM is implemented for image classification to visualize the predicted class scores and highlight the discriminative image features used by the CNN to classify the image. The implementation of CAM relies on performing Global Average Pooling (GAP) just before the final output layer. Using the above technique and the network architecture, the weights from the GAP layer can be used to highlight the important parts of the time series which led to the prediction. The obtained weights can then be used to visualize the explanation using the saliency mapping of the weight vector to the original time series.

MrSEQL-SM. Multi-resolution Symbolic Sequence Learner (MrSEQL) [6,11] classifier is an efficient TSC algorithm that trains a linear classification model. The algorithm transforms numeric time-series data into multiple symbolic representations of different domains such as SAX [13] in the time domain and SFA [22] in the frequency domain. The classifier selects the most important subsequences from the symbolic data which are then used as input features for training the SEQL classifier [11]. SEQL trains using logistic regression and outputs a linear model which is a set of weighted symbolic subsequences. For the SAX features which are in the time domain, saliency maps are then produced when these features and weights are mapped back to the original time series. This explanation produced in the form of a saliency map for MrSEQL with SAX features is called MrSEQL-SM [17].

2.4 Model-Agnostic Approaches

LIME. Local Interpretable Model-agnostic Explanations (LIME) [19] is a model-agnostic technique that explains the predictions of any classifier by approximating it locally with an interpretable model. In [19] the authors propose an implementation of LIME focused on training interpretable or local surrogate models to explain individual predictions. LIME examines how variations to the data fed into a black-box model, impact the model predictions. To achieve this, LIME perturbs the data and obtains black-box predictions for the new data points. Then, LIME trains an interpretable model on this perturbed dataset. The new samples are weighted according to their proximity to the instance of interest for which the explanation needs to be generated. This way LIME obtains the explanations for the instances locally and does not give a global approximation. LIME was previously implemented with text, image and tabular data [19]. For tabular data, variations of the data were produced by perturbing each feature individually. In the case of images, the variations are created by segmenting the image into *superpixels* which can be turned on or off with a user-defined colour. LIME can also be adapted for time series data as shown in [15,17]. Some of the key advantages of LIME are that it makes human-friendly and easily interpretable explanations and has local fidelity in terms of giving insight into explaining the black-box predictions locally [16]. LIME also has drawbacks, e.g., it samples data points using a Gaussian distribution which ignores feature correlation. There is also instability in the explanations produced, i.e., the explanations vary depending on some hyperparameters. An alternative to LIME is the Shapley value-based SHAP [14]. Even though SHAP gives benefits of local and global interpretability, it requires a lot of computation time since it is computing all possible feature permutations globally. Hence, LIME would have an advantage of speed when compared to SHAP. There is also no open implementation of SHAP for time series, hence we use LIME in this work.

ROCKET. RandOM Convolutional KErnal Transform (ROCKET) [4] is a classification method that transforms time series using random convolutional kernels (shape features) and trains a linear classifier using those transformed features. ROCKET can attain state-of-the-art accuracy using a fraction of the time as compared to other algorithms, including CNN. Since ROCKET uses a combination of shape features and numeric features - the proportion of positive values (ppv), it becomes difficult to obtain a saliency map directly from the linear model and we thus use LIME to obtain a post-hoc explanation for ROCKET, called ROCKET-LIME.

2.5   Evaluation Measures for Explanation Methods

According to [7], there are three main levels for the evaluation of interpretability - application grounded, human grounded, and function grounded. These vary in terms of complexity and the need according to different tasks. TSC explanation is aimed at focusing on the **discriminative** parts of the time series i.e., the parts important for classification. In TSC explanation, we want to evaluate explanations for individual predictions on the function level. There are several measures that can be used to judge how good an explanation method or explanation is [16]. Explanation methods have measures such as - *expressive power* in terms of the structure of the explanation generated by the model, *translucency* describing how much of the explanation method relies on looking into model parameters, *portability* describing the range of ML models that can implement this explanation method and the *algorithmic complexity* of the algorithm. Individual methods also possess an array of measures such as *accuracy* (how well the explanation reacts to unseen data), *fidelity* (how effectively the method estimates the prediction of black-box models), *consistency* (does the explanation vary between similar models or does it stay the same), *stability* (is a similar explanation generated on each iteration), *comprehensibility* (how well do humans understand the explanations), *certainty* (i.e. confidence of the model prediction), *degree of importance* (w.r.t the importance of features or parts of the explanation), *novelty* (is the explanation coming from a new distribution of the training data), and *coverage* in terms of the area covered.

Recent work [17] has used *informativeness* as an evaluation measure and the authors entail that if the explanation is truly informative, it should point out those parts of the time series that are most relevant for the classification decision. The authors highlight the discriminative parts of the time series by identifying a threshold k to find the parts where the weight vector belongs to the (100 - k) percentile discriminative weights. The authors have also made use of perturbation to provide evaluation for both single explanation methods as well as multiple explanation methods. In this work, we propose a novel methodology to calculate and compare informativeness. This extends the work of [17] and is a ranking-based methodology that uses perturbation to compute the ranks of multiple explanation methods over different referee classifiers and datasets. We choose informativeness over other evaluation measures because it helps in quantifying the evaluation for a single explanation and also gives an objective measure to perform a comparison of multiple explanation methods.

3   Proposed Methods

Here we discuss the technique used to perform the perturbation of the test set in order to evaluate the informativeness of a TSC explanation method. The perturbation process is then used for comparing different explanation methods based on their informativeness and for our ranking approach.

3.1   Explanation-Based Perturbation of Time Series

The main aim of a TSC explanation method is to emphasize those important regions of the time series that were most impactful for the classification decision. Hence, if an explanation is **informative**, it should point out those discriminative parts. In order to evaluate this, the discriminative regions of the time series test sets are perturbed to examine if a decrease in the classification accuracy is observed. The more informative the explanation is, the higher the expectation of a decrease in accuracy after perturbation based on this explanation method [17]. Here, we work with explanation methods that produce a saliency map for the time series. This information is stored as an array of positive weights w_t, one weight for each step in the time series having t steps. The discriminative weights are ranked through setting a threshold k ($0 \leq k \leq 100$) that is set at the (100 - k) percentile of the positive weight vector (w) that explains the time series [17]. Through this threshold, we can emphasize on the weights having the highest magnitude in the time series. For example, for $k = 10$, the focus with be on the top 10% of the highest weights coming from the explanation method. The time series is perturbed by adding Gaussian noise to its original signal. For a given time series represented by a vector x, the resulting perturbed vector is represented by $x_{perturbed}$ where the entire time series is perturbed and the distribution for the Gaussian noise is $N(\mu, \sigma^2)$, where μ is the mean of the distribution and σ is the magnitude of the noise.

$$x_{perturbed} = x + N(\mu, \sigma^2) \tag{1}$$

In this work, only a region is perturbed by adding noise based on the corresponding weights in the explanation vector. The rest of the time series remains unchanged. For the perturbation parameters we use $\mu = 0$ and $\sigma = 0.2 * range$. This effectively adds or subtracts about 20% of the magnitude range of values in that time series.

3.2   Calculating Informativeness as an Evaluation Metric

In order to quantitatively evaluate the informativeness of an explanation method, an experiment is proposed. Firstly, a time series classifier is trained using the original, non-perturbed training datasets as shown in Fig. 4. This classifier acts as an evaluation classifier or *referee classifier*. Thereafter, perturbed test datasets are created by adding noise to the discriminative parts of the time series. Multiple versions of the perturbed test datasets are obtained for multiple explanation methods, at the same threshold k ($0 \leq k \leq 100$). Each of these perturbed test datasets corresponds to an explanation or weight profile obtained from an explanation method.

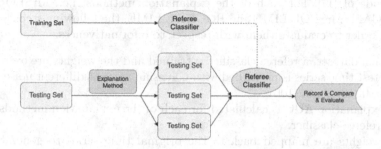

Fig. 4. Method of generating explanation-driven perturbed test sets and evaluating the explanation method through a referee classifier (reprinted from [17]).

If an explanation method is truly informative, the perturbation should impact the referee classifier more strongly than the other explanation methods. The informativeness of an explanation method is calculated by estimating the area under the explanation curve (AUC) described by accuracy at different perturbation levels k with the help of the trapezoidal rule. This metric is coined as an *explanation loss* or *eLoss* in the work [17] since a reduction of accuracy is observed after adding noise to the time series based on the given explanation method.

$$eLoss = \frac{1}{2}k \sum_{i=1}^{t}(acc_{i-1} + acc_i) \qquad (2)$$

Here, k represents the values of each step normalized in the range 0–1 where k = 0 corresponds to the original test dataset and the step k = 100 corresponds to perturbing the entire time series, t represents the number of steps in the time series (t = $\frac{100}{k}$) and acc_i represents the accuracy at step i. Here, we call the *eLoss* the **explanation AUC**, this is a numeric measure that varies between 0 and 1. The explanation methods are then compared using an *independent referee* classifier. In this work, we use three state of the art classifiers, MrSEQL, ROCKET and WEASEL and propose a new methodology to rank and compare explanations methods by aggregating over referees. The explanation methods are *ranked* based on their **explanation AUC** for each referee classifier. The lower the AUC, the higher the rank. Once the rank is calculated for an explanation method for one particular referee classifier, the overall rank is calculated by taking the average of all the obtained ranks across referees. The explanation method that ranks the highest is considered to be the most informative explanation method over the set of referees for that dataset. We provide more details on this strategy in the next section.

4   Experiments

Next, we discuss the steps required to generate the informativeness of each explanation method. We use the popular library sktime [24] and extend the open source code of [17]. For each of the explanation methods, i.e., MrSEQL-SM, ResNet-CAM, MrSEQL-LIME and ROCKET-LIME, the following steps are followed in order to evaluate them with respect to informativeness:

1. For each dataset, a referee classifier is trained and the weights are extracted.
2. Each test time series is perturbed with Gaussian noise at different noise levels k (i.e., 0, 10, 20, ..., 100).
3. The explanation AUC is calculated for each of the explanation methods with each referee classifier.
4. The weights are mapped back to the original time series to generate the saliency map for each method for each of the three datasets. The time taken to run and generate results for each explanation method is also recorded by using the `timeit` library.
5. The methods are then evaluated and ranked based on their informativeness using our proposed ranking-based methodology.

4.1   Perturbing and Measuring Metrics

An explanation method should point to discriminative parts of the time series if it is truly informative. If these discriminative parts are perturbed then a decrease in classification accuracy should be observed. Once the test datasets are perturbed, the new accuracy scores are generated and the explanation AUC is computed for each of the explanation methods with each referee classifier to computationally evaluate the usefulness of these explanation methods. Table 1 shows the accuracy at different noise levels k when using MrSEQL as a referee classifier, on the ROCKET-LIME explanation method, over the CMJ dataset. We note that the accuracy decreases as the noise levels increase from 10 to 100. As can also be seen in Fig. 5, this behaviour varies depending on the robustness to noise of the referee classifier. Table 2 shows the explanation AUC and the referee rank when using ROCKET as a referee classifier on the four explanation methods over the CMJ dataset.

Table 1. Accuracy for explanation ROCKET-LIME using MrSEQL as a referee classifier after adding Gaussian noise at levels k from 10–100 on the CMJ dataset.

Noise level	10	20	30	40	50	60	70	80	90	100
Accuracy	0.9609	0.9553	0.9553	0.9497	0.9497	0.9497	0.9441	0.9385	0.9385	0.9385

Table 2. Explanation AUC and rank for the explanation methods using ROCKET as a referee classifier over the CMJ dataset.

Dataset	Weights	Explanation AUC for referee ROCKET	Rank
CMJ	MrSEQL-SM	0.8874	2
CMJ	ResNet-CAM	0.9126	4
CMJ	MrSEQL-LIME	0.9115	3
CMJ	ROCKET-LIME	0.8866	1

4.2 Experimental Results and Evaluation

The four explanation methods are evaluated on the basis of their informativeness based on their ranking across the referee classifiers, over the datasets CMJ, Coffee and GunPoint. Due to the computational cost of LIME, MrSEQL-LIME is evaluated with only CMJ and GunPoint datasets whereas ROCKET-LIME is evaluated with the CMJ dataset only.

Accuracy. Figure 5 shows the accuracy curve for the CMJ dataset after Gaussian noise is added to the time series. This is shown for all the four explanation methods and the three referee classifiers. It can be seen that as the noise levels increase from zero to a hundred, a dip in referee accuracy is seen for all the explanation methods. This supports the fact that performing perturbation decreases the accuracy of the referees.

In order to compare the explanation methods against each other based on the accuracy curve, the accuracy curves are aggregated to see which method is the most informative. The lower curve indicates that performing perturbation decreases the accuracy of the explanation method more. This indicates that the explanation method is more informative. Figure 6 shows the comparison of the

Table 3. Explanation AUC obtained for the four explanation methods and referee classifiers. In bold is the lowest AUC over explanations, for a given referee, which results in rank 1 for that explanation method and referee.

Dataset	Explanation method	MrSEQL	ROCKET	WEASEL
CMJ	MrSEQL-SM	**0.9441**	0.8874	**0.6575**
	ResNet-CAM	0.9453	0.9126	0.6793
	MrSEQL-LIME	**0.9441**	0.9115	0.6933
	ROCKET-LIME	0.9492	**0.8866**	0.7039
Coffee	MrSEQL-SM	**0.9625**	**1.000**	0.9804
	ResNet-CAM	0.9696	**1.000**	**0.9696**
GunPoint	MrSEQL-SM	**0.9477**	**0.7137**	0.5440
	ResNet-CAM	0.9610	0.7350	**0.5280**
	MrSEQL-LIME	0.9677	0.7637	0.5727

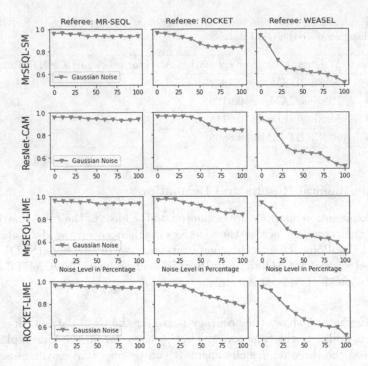

Fig. 5. The change in accuracy when perturbation is performed by adding Gaussian noise to the test time series for each explanation method from (top to down) with the three referee classifiers from (left to right) on the CMJ dataset.

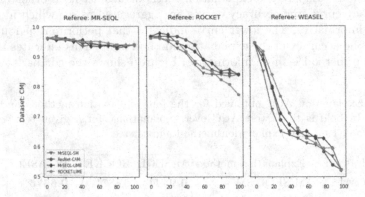

Fig. 6. Comparison of accuracy after perturbation with Gaussian noise for MrSEQL-SM, ResNet-CAM, MrSEQL-LIME and ROCKET-LIME using the CMJ dataset and the referee classifiers, MrSEQL, ROCKET and WEASEL. The lower curve indicates more impact of the explanation method on the referee classification accuracy.

accuracy curves for all four explanation methods on the CMJ dataset. It can be seen that although there is an overlap between the curves, MrSEQL-SM shown by the red curve is slightly more informative as compared to the other methods.

Explanation AUC. Table 3 represents the explanation AUC obtained for each of the datasets and the explanation methods across the referee classifiers. We observe that the explanation AUC varies across the three classifiers. The lower explanation AUC value indicates a higher referee rank contributing towards higher informativeness.

Informativeness. The explanation methods are *ranked* and evaluated based on their explanation AUC for each classifier. Then the overall rank is calculated as the average-rank by aggregating over the referees as shown in Table 4. The explanation method that ranks the highest is taken as the most informative explanation method.

Table 4. Ranking of explanation methods based on their informativeness.

Dataset	Explanation method	MrSEQL	ROCKET	WEASEL	Average rank
CMJ	MrSEQL-SM	1.00	2.00	1.00	1.33
	MrSEQL-LIME	1.00	3.00	3.00	2.33
	ResNet-CAM	2.00	4.00	2.00	2.67
	ROCKET-LIME	3.00	1.00	4.00	2.67
Coffee	MrSEQL-SM	1.00	1.00	2.00	1.33
	ResNet-CAM	2.00	1.00	1.00	1.33
GunPoint	MrSEQL-SM	1.00	1.00	2.00	1.33
	ResNet-CAM	2.00	2.00	1.00	1.67
	MrSEQL-LIME	3.00	3.00	3.00	3.00

We make the following observations with regards to the average rank of explanation methods for each dataset:

- **CMJ:** MrSEQL-SM has the highest average rank and is thus the most informative followed by MrSEQL-LIME, ResNet-CAM and ROCKET-LIME. Both ResNet-CAM and ROCKET-LIME seem to be equally informative due to a similar average rank.
- **Coffee:** Both MrSEQL-SM and ResNet-CAM show the same average rank and hence are equally informative.
- **GunPoint:** MrSEQL-SM is the most informative followed by ResNet-CAM and then MrSEQL-LIME.

It is also important to note that the ranks vary across different referee classifiers and the referee classifier contributes towards the informativeness computation of the explanation methods. Even though MrSEQL-SM performs well with

MrSEQL and WEASEL as referee classifiers for the CMJ dataset, it ranks second in the case of ROCKET as a referee classifier. This is also seen for ROCKET-LIME as it ranks first when trained with ROCKET itself as a referee classifier but not in other cases. Therefore, to obtain an aggregate behaviour of each explanation method over referees, an average rank is computed. MrSEQL-SM has the highest average rank followed by ResNet-CAM, MrSEQL-LIME and finally ROCKET-LIME. It is also important to note that these ranks depend on the problem statement and the dataset, and can be different for different datasets. Further work is also needed to evaluate MrSEQL-LIME with Coffee dataset and ROCKET-LIME with GunPoint and Coffee datasets.

Runtime Analysis. The runtime of each explanation method to train a classifier, return the weights and plot the explanation in the form of a saliency map is calculated and displayed in Table 5. For each dataset, the run time is observed when performing the experiment for each of the explanation methods.

Table 5. Time (seconds) for model training, getting weights and getting the explanation for all the explanation methods for each dataset.

Dataset	Method	TrainingTime	GettingWeights	GettingExplanation	Total(sec)
CMJ	MrSEQL-SM	362.74	134.57	4.17	501.48
	ResNet-CAM	6.10	5.47	7.62	19.19
	MrSEQL-LIME	178.28	4438.53	4.75	4621.56
	ROCKET-LIME	33.26	3962.98	6.28	4002.53
Coffee	MrSEQL-SM	5.37	1.57	2.12	9.07
	ResNet-CAM	1.86	1.34	0.73	3.93
GunPoint	MrSEQL-SM	4.61	2.73	2.35	9.70
	ResNet-CAM	1.75	3.81	0.84	6.41
	MrSEQL-LIME	4.13	465.38	3.21	472.73

The following points summarize the findings for each dataset:

- **CMJ:** ResNet-CAM is the fastest to reproduce the results since a pre-trained model is used for training, otherwise ROCKET would be the fastest. After ResNet-CAM, we have MrSEQL-SM followed by ROCKET-LIME and then finally MrSEQL-LIME.
- **Coffee:** ResNet-CAM is somewhat faster than MrSEQL-SM in getting the weights and the explanation. Even if the classifier is fast to train, adding an explanation with LIME makes the explanation step slow.
- **GunPoint:** MrSEQL-LIME is computationally expensive as opposed to ResNet-CAM and MrSEQL-SM.

We note that model-specific approaches such as MrSEQL-SM and ResNet-CAM are much faster than model agnostic approaches involving LIME, i.e., MrSEQL-LIME and ROCKET-LIME. Hence, even though ROCKET is an extremely fast classification method, its computational cost increases when it is combined with LIME to obtain an explanation.

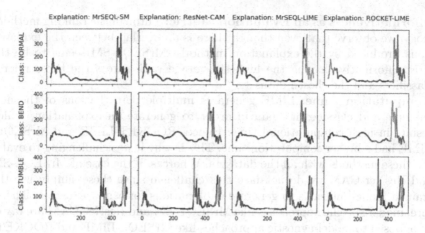

Fig. 7. Saliency maps produced by MrSEQL-SM, ResNet-CAM, MrSEQL-LIME and ROCKET-LIME explanation methods for an example time series from the three classes of the CMJ dataset.

Visualizing Saliency Mappings. The weights extracted for each of the explanation methods are mapped back to the time series in order to visualize them with the help of a saliency map. The most discriminative regions of the time series are highlighted in red by the explanation methods whereas the least discriminative regions are highlighted in blue by the explanation method on a scale of 0–100. Saliency maps help us validate the informativeness of the explanation methods. Each explanation method is compared with one another based on the generated saliency. It is clear from the figures shown in the sections below that all the methods give different explanations highlighting the importance of an objective evaluation approach. The CMJ dataset contains three classes - NORMAL, BEND and STUMBLE. Figure 7 shows the saliency maps generated by MrSEQL-SM, ResNet-CAM, MrSEQL-LIME and ROCKET-LIME for the CMJ dataset. It can be seen that each explanation method highlights a different region to be most informative. MrSEQL-SM appears to be the most informative since it clearly highlights the low-middle parts of the class NORMAL, the hump-middle part of the class BEND and the high peak part of the class STUMBLE (please refer to [11] for details on discriminative regions in this dataset). MrSEQL-LIME and ROCKET-LIME also highlight similar regions however, the explanations produced by MrSEQL-LIME is more similar to MrSEQL-SM than ROCKET-LIME. On the other hand, ResNet-CAM does not clearly highlight known discriminative parts in the time series of this dataset.

Discussion. From the previous experiments we observe that explanation methods can indeed be quantitatively compared using the notion of informativeness based on ranking. The key takeaways from this work are summarized below.

- **Informativeness as an Evaluation Metric:** Through the ranking methodology, we observe that even though there is an overlap between the explanations produced by the explanation methods, MrSEQL-SM seems to be the most informative having the highest average rank across the three referee classifiers and the chosen datasets.
- **Computation Time:** LIME generates multiple perturbations of the new example and classifies it again in order to generate an explanation, which results in high computational time. Hence, ROCKET-LIME and MrSEQL-LIME had a higher computation time. This is why it was challenging to evaluate these methods with all the datasets. Whereas in the case of MrSEQL-SM and ResNet-CAM we do not face this challenge since these simply use the trained model internals to generate explanations for a new example. Therefore, model-specific approaches like MrSEQL-SM and ResNet-CAM are faster as opposed to model-agnostic approaches like MrSEQL-LIME and ROCKET-LIME.
- **Impact of Referee Classifier:** The referee classifier can impact the classification accuracy and the explanation AUC of the explanation methods. We can also observe from Fig. 6 that ROCKET and WEASEL appear to be more sensitive to the noise added during perturbation and show a significant reduction of accuracy as the amount of Gaussian noise added increases. This is not the case for MrSEQL as a referee classifier since the reduction is not that significant. Note that here MrSEQL only uses SAX features (in the time domain), while WEASEL uses SFA features (in the frequency domain), and ROCKET uses a mix of features in the time domain (i.e., convolution kernels) and features similar to the frequency domain features (i.e., dilation).
- **Saliency Mappings:.** Saliency maps can yield an accurate visual representation of what parts of the time series are considered important by the explanation method. This not only cross evaluates the ranking methodology but also represents the vector of weights in a visual manner.

5   Conclusion

This work aimed to quantitatively evaluate the informativeness of different model-specific as well as model-agnostic explanation methods for TSC. Through experimental results, we showed that TSC explanation methods can be evaluated and ranked based on their informativeness and that saliency-based visualizations support the results attained. Our simple ranking-over-referees technique can be implemented for practical applications in order to evaluate current TSC explanation methods or understand the classification decision-making process of TSC algorithms. In this work, four explanation methods are explored on three datasets, however, this technique can be adopted and expanded to evaluate other explanation methods and datasets based on the needs of a given problem statement. For future work we will extend the study of perturbation approaches, extend the set of referees and apply this methodology to more datasets that have available explanation ground truth. Given the fast growth of XAI and the amount of new methods proposed for explaining classifiers, we consider that having an effective methodology to objectively evaluate and compare these methods is very important to make sure that real progress is made and that the new explanation methods are actually useful.

Acknowledgments. This publication has emanated from research supported in part by a grant from Science Foundation Ireland through the SFI Centre for Research Training in Machine Learning (18/CRT/6183), the Insight Centre for Data Analytics (12/RC/2289_P2) and the VistaMilk SFI Research Centre (SFI/16/RC/3835). For the purpose of Open Access, the author has applied a CC BY public copyright licence to any Author Accepted Manuscript version arising from this submission. The authors would like to thank the reviewers for their constructive feedback.

References

1. Apley, D.W., Zhu, J.: Visualizing the effects of predictor variables in black box supervised learning models. J. R. Stat. Soc. Ser. B Stat. Methodol. **82**(4), 1059–1086 (2020). https://doi.org/10.1111/rssb.12377
2. Bagnall, A., Flynn, M., Large, J., Lines, J., Middlehurst, M.: A tale of two toolkits, report the third: on the usage and performance of HIVE-COTE v1.0 (2020). http://arxiv.org/abs/2004.06069
3. Bagnall, A., Lines, J., Bostrom, A., Large, J., Keogh, E.: The great time series classification bake off: a review and experimental evaluation of recent algorithmic advances. Data Min. Knowl. Disc. **31**(3), 606–660 (2016). https://doi.org/10.1007/s10618-016-0483-9
4. Dempster, A., Petitjean, F., Webb, G.I.: ROCKET: exceptionally fast and accurate time series classification using random convolutional kernels. DAMI. https://link.springer.com/article/10.1007/s10618-020-00701-z
5. Deng, H., Runger, G., Tuv, E., Vladimir, M.: A time series forest for classification and feature extraction. Inf. Sci. **239**, 142–153 (2013)
6. Dhariyal, B., Nguyen, T.L., Gsponer, S., Ifrim, G.: An examination of the state-of-the-art for multivariate time series classification. In: ICDMW (2020)
7. Doshi-Velez, F., Kim, B.: Towards a rigorous science of interpretable machine learning (2017)
8. Du, M., Liu, N., Hu, X.: Techniques for interpretable machine learning (2019)
9. Ismail Fawaz, H., Forestier, G., Weber, J., Idoumghar, L., Muller, P.-A.: Deep learning for time series classification: a review. Data Min. Knowl. Disc. **33**(4), 917–963 (2019). https://doi.org/10.1007/s10618-019-00619-1
10. Kim, B., Khanna, R., Koyejo, O.O.: Examples are not enough, learn to criticize! Criticism for interpretability. In: NeurIPS, vol. 29, pp. 2280–2288. Curran Associates, Inc. (2016)
11. Le Nguyen, T., Gsponer, S., Ilie, I., O'Reilly, M., Ifrim, G.: Interpretable time series classification using linear models and multi-resolution multi-domain symbolic representations. Data Min. Knowl. Disc. **33**(4), 1183–1222 (2019). https://doi.org/10.1007/s10618-019-00633-3
12. Lei, Y., Wu, Z.: Time series classification based on statistical features. EURASIP J. Wirel. Commun. Netw. **2020**(1), 1–13 (2020). https://doi.org/10.1186/s13638-020-1661-4

13. Lin, J., Keogh, E., Wei, L., Lonardi, S.: Experiencing SAX: a novel symbolic representation of time series. DAMI **15**(2), 107–144 (2007)
14. Lundberg, S., Lee, S.I.: A unified approach to interpreting model predictions (2017)
15. Metzenthen, E.: Lime for time code repository. https://github.com/emanuel-metzenthin/Lime-For-Time/blob/master/demo/LIME-Pipeline.ipynb
16. Molnar, C.: Interpretable machine learning. https://christophm.github.io/interpretable-ml-book/
17. Nguyen, T.T., Le Nguyen, T., Ifrim, G.: A model-agnostic approach to quantifying the informativeness of explanation methods for time series classification. In: Lemaire, V., Malinowski, S., Bagnall, A., Guyet, T., Tavenard, R., Ifrim, G. (eds.) AALTD 2020. LNCS (LNAI), vol. 12588, pp. 77–94. Springer, Cham (2020). https://doi.org/10.1007/978-3-030-65742-0_6
18. Ozyegen, O., Ilic, I., Cevik, M.: Evaluation of local explanation methods for multivariate time series forecasting, pp. 1–13 (2020). http://arxiv.org/abs/2009.09092
19. Ribeiro, M.T., Singh, S., Guestrin, C.: "Why should i trust you?" explaining the predictions of any classifier. In: KDD, pp. 1135–1144 (2016)
20. Santos, T., Kern, R.: A literature survey of early time series classification and deep learning. In: CEUR Workshop Proceedings, vol. 1793 (2017)
21. Schäfer, P.: The BOSS is concerned with time series classification in the presence of noise. DAMI **29**(6), 1505–1530 (2015). https://doi.org/10.1007/s10618-014-0377-7
22. Schäfer, P., Högqvist, M.: SFA: a symbolic Fourier approximation and index for similarity search in high dimensional datasets. In: EDBT, pp. 516–527 (2012)
23. Schäfer, P., Leser, U.: Fast and accurate time series classification with WEASEL. In: CIKM, pp. 637–646 (2017)
24. Turing, A.: Sktime specifications. https://www.turing.ac.uk/research/research-projects/sktime-toolbox-data-science-time-series
25. Ye, L., Keogh, E.: Time series shapelets: a novel technique that allows accurate, interpretable and fast classification. DAMI **22**(1–2), 149–182 (2011)
26. Zhou, B., Khosla, A., Lapedriza, A., Oliva, A., Torralba, A.: Learning deep features for discriminative localization (2015)

State Space Approximation of Gaussian Processes for Time Series Forecasting

Alessio Benavoli[1]([✉]) [ID] and Giorgio Corani[2] [ID]

[1] School of Computer Science and Statistics (SCSS), Trinity College Dublin,
Dublin, Ireland
`alessio.benavoli@tcd.ie`
[2] Istituto Dalle Molle di Studi sull'Intelligenza Artificiale (IDSIA), USI - SUPSI
Lugano, Lugano, Switzerland
`giorgio.corani@idsia.ch`

Abstract. Gaussian Processes (GPs), with a complex enough additive kernel, provide competitive results in time series forecasting compared to state-of-the-art approaches (arima, ETS) provided that: (i) during training the unnecessary components of the kernel are made irrelevant by automatic relevance determination; (ii) priors are assigned to each hyperparameter. However, GPs computational complexity grows cubically in time and quadratically in memory with the number of observations. The state space (SS) approximation of GPs allows to compute GPs based inferences with linear complexity. In this paper, we apply the SS representation to time series forecasting showing that SS models provide a performance comparable with that of full GP and better than state-of-the-art models (arima, ETS). Moreover, the SS representation allows us to derive new models by, for instance, combining ETS with kernels.

Keywords: Time series forecasting · Gaussian Process · State space approximation

1 Introduction

Gaussian Processes (GPs) [15] are a powerful tool for modeling correlated observations, including time series. GPs have been used for the analysis of astronomical time series (see [4] and the references therein), forecasting of electric load [12] and analysis of correlated and irregularly-sampled time series [16].

A kernel composition specific for time series has been recently proposed [3]. It contains linear trend, periodic patterns, and other flexible kernel for modeling the non-linear trend. By setting priors on the hyperparameters, which keep the inference within a reasonable range even on short time series, the GP yields very accurate forecasts, outperforming the traditional time series models.

Note that the above GP based model is a type of Generalised Additive Model (GAM) [26]. However, contrarily to traditional GAMs, it uses different nonparametric components for the periodic and non-linear terms, and it is estimated in a fully Bayesian way (that is, without backfitting).

© Springer Nature Switzerland AG 2021
V. Lemaire et al. (Eds.): AALTD 2021, LNAI 13114, pp. 21–35, 2021.
https://doi.org/10.1007/978-3-030-91445-5_2

Yet, GPs have computational complexity $O(n^3)$ and storage demands of $O(n^2)$; hence, they are not suitable for large datasets. Several approximations have been proposed to reduce their computational complexity to $O(n)$, such as sparse approximations based on inducing points [1,6,7,14,19,20,24], which however add additional hyperparameters.

In the case of time series, it is possible to represent the full GP as a State Space model, without the need for any additional hyperparameter [2,11,13,17, 18,22] and with $O(n)$ complexity.

We focus on the SS representation of the GP and we provide the following contributions. We discuss how to represent the model of [3] as a SS model, obtaining almost identical results on the time series of the M3 competition.

We also apply the GP model of [3] to very long time series, thanks to the SS representation. Also in this case we obtain positive results w.r.t the competitors.

Moreover, once the covariance functions of the Gaussian are represented in the SS framework, they can be combined with the existing SS models. This opens up the possibility of developing novel time series models. As a proof of concept, we consider a traditional state-space model (additive exponential smoothing) and we replace its seasonal component with the SS representation of the periodic kernel of the GP. We obtain a less parameterized model, which has higher accuracy on the time series of the M3 competition. The resulting model is also more flexible; for instance, it could be easily extended to manage time series containing multiple seasonal patterns, unlike the traditional exponential smoothing.

2 Background

In the following section, we provide a background on (i) Gaussian Processes; (ii) State Space models; (iii) the State Space representation of Gaussian Processes.

2.1 Gaussian Process

We consider the regression model

$$y = f(\mathbf{x}) + v, \tag{1}$$

where $\mathbf{x} \in \mathbb{R}^p$, $f : \mathbb{R}^p \to \mathbb{R}$ and $v \sim N(0, s_v^2)$ is the noise. Our goal is to estimate f given the training data $\mathcal{D} = \{(\mathbf{x}_i, y_i), \ i = 1, \ldots, n\}$. In GP regression, we place a GP prior on the unknown f, $f \sim GP(0, k_\theta)$,[1] and calculate the posterior distribution of f given the data $\mathcal{D}$. We then employ this posterior to make inferences about f.

In particular, we are interested in predictive inferences. Based on the training data $X^T = [\mathbf{x}_1, \ldots, \mathbf{x}_n]$, $\mathbf{y} = [y_1, \ldots, y_n]^T$, and given m test inputs $(X^*)^T = [\mathbf{x}_1^*, \ldots, \mathbf{x}_m^*]$, we aim to find the posterior distribution of $\mathbf{f}^* =$

---

[1] A GP prior with zero mean function and covariance function $k_\theta : \mathbb{R}^p \times \mathbb{R}^p \to \mathbb{R}^+$, which depends on a vector of hyperparameters θ.

$[f(\mathbf{x}_1^*), \ldots, f(\mathbf{x}_m^*)]^T$. From (1) and the properties of the Gaussian distribution,[2] the posterior distribution of $\mathbf{f}^*$ is Gaussian [15, Sec. 2.2]:

$$p(\mathbf{f}^*|X^*, X, \mathbf{y}, \boldsymbol{\theta}) = N(\mathbf{f}^*; \hat{\boldsymbol{\mu}}_\theta(X^*|X, \mathbf{y}), \hat{K}_\theta(X^*, X^*|X)), \qquad (2)$$

with mean and covariance given by:

$$\hat{\boldsymbol{\mu}}_\theta(\mathbf{f}^*|X, \mathbf{y}) = K_\theta(X^*, X)(K_\theta(X, X))^{-1}\mathbf{y},$$
$$\hat{K}_\theta(X^*, X^*|X) = K_\theta(X^*, X^*) - K_\theta(X^*, X)(K_\theta(X, X))^{-1}K_\theta(X, X^*). \qquad (3)$$

In GPs, the kernel defines the Covariance Function (CF) between any two function values: $Cov(f(\mathbf{x}), f(\mathbf{x}^*)) = k_\theta(\mathbf{x}, \mathbf{x}^*)$. Common kernels are the White Noise (WN), the Linear (LIN), the Matern 3/2 (MAT32), the Matern 5/2 (MAT52), the Squared Exponential (RBF), the Cosine (COS) and the Periodic (PER). Hereafter, we provide the expressions of these kernels for $p = 1$, which is the case of time series; see instead [15] for generalizations:

WN: $k_\theta(x_1, x_2) = s_v^2 \delta_{x_1, x_2}$

LIN: $k_\theta(x_1, x_2) = s_b^2 + s_l^2 x_1 x_2$

MAT32: $k_\theta(x_1, x_2) = s_e^2 \left(1 + \frac{\sqrt{3}|x_1 - x_2|}{\ell_e}\right) \exp\left(-\frac{\sqrt{3}|x_1 - x_2|}{\ell_e}\right)$

MAT52: $k_\theta(x_1, x_2) = s_e^2 \left(1 + \frac{\sqrt{5}|x_1 - x_2|}{\ell_e} + \frac{5(x_1 - x_2)^2}{3\ell_e^2}\right) \exp\left(-\frac{\sqrt{5}|x_1 - x_2|}{\ell_e}\right)$

RBF: $k_\theta(x_1, x_2) = s_r^2 \exp\left(-\frac{(x_1 - x_2)^2}{2\ell_r^2}\right)$

COS: $k_\theta(x_1, x_2) = s_c^2 \cos\left(\frac{x_1 - x_2}{\tau}\right)$

PER: $k_\theta(x_1, x_2) = s_p^2 \exp\left(-\frac{(2\sin^2(\pi|x_1 - x_2|/p_e)}{\ell_p^2}\right)$

where δ_{x_1, x_2} is the Kronecker delta, which equals one when $x_1 = x_2$ and zero otherwise. The hyperparameters are the variances $s_v^2, s_l^2, s_e^2, s_r^2, s_c^2, s_p^2 > 0$, the lengthscales $\ell_r, \ell_e, \ell_p, \tau > 0$ and the period p_e.

Selecting a kernel, or a combination of kernels, to determine the structure of the covariance is a crucial factor governing the performance of a GP model. Spectral mixture kernels (SM) [25] have been devised to overcome this issue thanks to their property of being able to approximate any stationary kernel.[3] SM define a covariance kernel by taking the inverse Fourier transform of a weighted sum of different shifts of a probability density. In the original formulation [25],

---

[2] In this work, we include the additive noise v into the kernel by adding a White noise kernel term.

[3] A stationary kernel is one which is translation invariant: $k_\theta(x_1, x_2)$ depends only on $x_1 - x_2$, like for instance the Matern and RBF kernels.

the authors considered a Gaussian PDF, resulting into a covariance kernel which is the sum of the RBF×COS kernels, so each term in the sum is equal to:

$$\text{SM}_i:\ k_\theta(x_1, x_2) = s_{m_i}^2 \exp\left(-\frac{(x_1 - x_2)^2}{2\ell_{m_i}^2}\right) \cos\left(\frac{x_1 - x_2}{\tau_{m_i}}\right),$$

with hyperparameters s_{m_i}, ℓ_{m_i} and τ_{m_i}.

Learning the Hyperparameters. We denote by $\boldsymbol{\theta}$ the vector containing all the kernels' hyperparameters. In practical application of GPs, $\boldsymbol{\theta}$ have to be selected. We use Bayesian model selection to consistently set such parameters. Variances and lengthscales are non-negative hyperparameters, to which we assign log-normal priors (later we show how we define the priors). We then compute the maximum a-posteriori (MAP) estimate of $\boldsymbol{\theta}$, that is we maximize w.r.t. $\boldsymbol{\theta}$ the joint marginal probability $p(\mathbf{y}, \boldsymbol{\theta})$, which is the product of the prior $p(\boldsymbol{\theta})$ and the marginal likelihood [15, Ch.2]:

$$p(\mathbf{y}|X, \boldsymbol{\theta}) = N(\mathbf{y}; 0, K_\theta(X, X)). \tag{4}$$

Usually $\boldsymbol{\theta}$ is selected by maximizing the marginal likelihood of Eq. (4). Yet, better estimates can be obtained by assigning prior to the hyperparameters and then performing MAP estimation. The MAP approach yields reliably estimates also on short time series, as pointed out by [3], in which it is also proposed a methodology to define such priors.

2.2   State Space Models

Consider the following stochastic continuous time-variant (LTV) State Space (SS) model [10]

$$\begin{cases} d\mathbf{f}(t) = \mathbf{F}(t)\,\mathbf{f}(t)dt + \mathbf{L}(t)\,dw(t), \\ y(t_k) = \mathbf{C}(t_k)\,\mathbf{f}(t_k), \end{cases} \tag{5}$$

where $\mathbf{f}(t) = [f_1(t), \ldots, f_m(t)]^T$ is the state vector,[4] $y(t_k)$ is the measurement at time t_k, $\mathbf{F}(t), \mathbf{C}(t), \mathbf{L}(t)$ are known matrices of appropriate dimensions and $w(t)$ is a one-dimensional Wiener noise process with intensity $q(t)$. We further assume that the initial state $\mathbf{f}(t_0)$ and $w(t)$ are independent for each $t \geq t_0$. The solution of the stochastic differential equation in (5) is [10]:

$$\mathbf{f}(t_k) = \boldsymbol{\psi}(t_k, t_0)\,\mathbf{f}(t_0) + \int_{t_0}^{t_k} \boldsymbol{\psi}(t_k, \tau)\mathbf{L}(\tau)\,dw(\tau), \tag{6}$$

---

[4] m is a latent dimension which defines the dimension of the state space. The state is a function of *tim*.

with $\psi(t_k, t_0) = \exp(\int_{t_0}^{t_k} \mathbf{F}(t)dt)$ is the state transition matrix, which is computed as a matrix exponential.[5] Assuming that $E[\mathbf{f}(t_0)] = \mathbf{0}$, then it can be easily proven that the vector of observations $[y(t_1), y(t_2), \ldots, y(t_n)]^T$ is Gaussian distributed with zero mean and covariance matrix whose elements are given by:

$$E[y(t_i)y(t_j)] = \mathbf{C}(t_i)\psi(t_i, t_0)E[\mathbf{f}(t_0)\mathbf{f}^T(t_0)](\mathbf{C}(t_j)\psi(t_j, t_0))^T$$
$$+ \int_{t_0}^{\min(t_i, t_j)} h(t_i, u)h(t_j, u)q(u)du \qquad (7)$$

where we have exploited the fact that $E[dw(u)dw(v)] = q(u)\delta(u-v)dudv$ [10] and defined $h(t_1, t_2) = \mathbf{C}(t_1)\psi(t_1, t_2)\mathbf{L}(\tau)$.

In SS models, one aims to estimate the states $\mathbf{f}(t_1), \ldots, \mathbf{f}(t_n)$ given the observations $y(t_1), \ldots, y(t_n)$ and the initial condition. There are in particular two problems of interest: (i) *filtering* whose aim is to compute $p(\mathbf{f}(t_k)|y(t_1), \ldots, y(t_k))$ for every t_k; (ii) *smoothing* whose aim is to compute $p(\mathbf{f}(t_k)|y(t_1), \ldots, y(t_n))$ for every t_k. For stochastic LTV systems, filtering and smoothing can be solved exactly using the Kalman Filter (KF) and the Rauch-Tung-Striebel smoother [10] with complexity $\mathcal{O}(n)$.

2.3   SS Models Representation of GPs

When the GP has one-dimensional input, it is possible to represent (or approximate) the GP with a SS model. The advantage of the SS representation is that estimates and inferences can be computed with complexity $\mathcal{O}(n)$. In practice, one has to find a SS whose covariance matrix (7) coincides (or approximates) that of the GP. This provides the SS representation of the GP, which then allows us to estimate $\mathbf{f}(t_k)$ given data $\{y(t_1), \ldots, y(t_n)\}$ using the KF and the Rauch-Tung-Striebel smoother (with complexity $\mathcal{O}(n)$). This can be obtained as follows:

1. Discretize the continuous-time SS to obtain a discrete-time SS (this step basically consists on applying (6)):

$$\begin{cases} \mathbf{f}(t_k) = \psi(t_k, t_{k-1})\mathbf{f}(t_{k-1}) + \boldsymbol{\nu}(t_{k-1}), \\ y(t_k) = \mathbf{C}(t_k)\mathbf{f}(t_k), \end{cases} \qquad (8)$$

where $\boldsymbol{\nu}(t_{k-1}) = \int_{t_{k-1}}^{t_k} \psi(t_k, \tau)\mathbf{L}\, dw(\tau)$.

2. Compute the probability density function (PDF) $p(\mathbf{f}(t_k)|y(t_1), \ldots, y(t_k))$, which is Gaussian. The mean and covariance matrix of this Gaussian PDF can be computed efficiently by using the KF.

3. Compute the Gaussian posterior PDF $p(\mathbf{x}(t_k)|y(t_1), \ldots, y(t_n))$ – the mean and covariance matrix of this PDF can be computed very efficiently by using the Rauch-Tung-Striebel smoother. This step returns the estimates of the state given all observations.

---

[5] The matrix exponential is $e^A = I + A + A^2/2! + A^3/3! + \ldots$ and, for many matrices A, it can be computed analytically.

4. To estimate the hyperparameters of the CF, we can perform MAP (as for GPs). Note that, the marginal likelihood of the SS model can be computed efficiently by the KF.

State Space Representation of Covariance Functions. The time continuous SS representation of the covariance functions of Sect. 2.1 is given in Table 1. Such representations do not include the variance scaling parameter that multiplies the CF; it can be however included in the SS model by rescaling either the stochastic forcing term or the initial condition (for SS without forcing term).

Table 1. SS representation of the CFs. When the distribution of the initial state is not provided, it is assumed to be equal to zero. The intensity of the Wiener process w is assumed to be $q = 1$.

WN	$\begin{cases} \frac{df}{dt}(t) = \frac{dw}{dt}(t) \\ y(t_k) = f(t_k) \end{cases}$
LIN	$\begin{cases} \frac{df_1}{dt}(t) = f_2(t) \\ \frac{df_2}{dt}(t) = 0 \\ y(t_k) = f_1(t_k) \end{cases}$ $\quad \begin{bmatrix} f_1(t_0) \\ f_2(t_0) \end{bmatrix} \sim \mathcal{N}\left(\begin{bmatrix} 0 \\ 0 \end{bmatrix}, \begin{bmatrix} s_b^2 & 0 \\ 0 & s_l^2 \end{bmatrix} \right)$
MAT32	$\begin{cases} \frac{df_1}{dt}(t) = f_2(t) \\ \frac{df_2}{dt}(t) = -\frac{3}{\ell^2} f_1(t) - \frac{2\sqrt{3}}{\ell} f_2(t) + \frac{12\sqrt{3}}{\ell^3} \frac{dw}{dt}(t) \\ y(t_k) = f_1(t_k) \end{cases}$
MAT52	$\begin{cases} \frac{df_1}{dt}(t) = f_2(t) \\ \frac{df_2}{dt}(t) = f_3(t) \\ \frac{df_3}{dt}(t) = -\frac{3\sqrt{5}}{\ell} f_1(t) - \frac{15}{\ell^2} f_2(t) - \frac{3\sqrt{5}}{\ell} f_3(t) + \frac{400\sqrt{5}}{3\ell^5} \frac{dw}{dt}(t) \\ y(t_k) = f_1(t_k) \end{cases}$
COS	$\begin{cases} \frac{df_1}{dt}(t) = \frac{1}{\tau} f_2(t) \\ \frac{df_2}{dt}(t) = -\frac{1}{\tau} f_1(t) \\ y(t_k) = f_1(t_k) \end{cases}$ $\quad \begin{bmatrix} f_1(t_0) \\ f_2(t_0) \end{bmatrix} \sim \mathcal{N}\left(\begin{bmatrix} 0 \\ 0 \end{bmatrix}, \begin{bmatrix} 1 & 0 \\ 0 & 1 \end{bmatrix} \right)$

Representing Compositions of Covariance Functions. Additive combination of covariance functions can be represented by stacking SS models; this is called *cascade composition* [17]. For instance, the SS model corresponding to WN+LIN is:

$$\begin{cases} \frac{df_1}{dt}(t) = \frac{dw}{dt}(t) \\ \frac{df_2}{dt}(t) = f_3(t) \\ \frac{df_3}{dt}(t) = 0 \\ y(t_k) = f_1(t_k) + f_2(t_k) \end{cases} \qquad \begin{bmatrix} f_2(t_0) \\ f_3(t_0) \end{bmatrix} \sim \mathcal{N}\left(\begin{bmatrix} 0 \\ 0 \end{bmatrix}, \begin{bmatrix} s_b^2 & 0 \\ 0 & s_l^2 \end{bmatrix} \right).$$

Multiplicative composition of covariance functions can be obtained via *parallel composition* [17] of SS models. For instance, the COS × MAT32 kernel is represented as:

$$\begin{cases} \frac{df_1}{dt}(t) = \omega f_2(t) + f_3(t) \\ \frac{df_2}{dt}(t) = -\omega f_1(t) + f_4(t) \\ \frac{df_3}{dt}(t) = -\frac{3}{\ell^2} f_1(t) - \frac{2\sqrt{3}}{\ell} f_3(t) + \omega f_4(t) + \frac{12\sqrt{3}}{\ell^3} \frac{dw_1}{dt}(t) \\ \frac{df_4}{dt}(t) = -\frac{3}{\ell^2} f_2(t) - \omega f_3(t) - \frac{2\sqrt{3}}{\ell} f_4(t) + \frac{12\sqrt{3}}{\ell^3} \frac{dw_2}{dt}(t) \\ y(t_k) = f_1(t_k) \end{cases}$$

The RBF and PER kernel do not admit an exact SS representation; for this reason, they are not shown in Table 1. However, an approximated SS representation can be given. The PER kernel can be approximated as the sum of different Cosine covariance functions (COS + COS + ... + COS), with a suitable choice of their lengthscales (defined using a Fourier series expansion of the PER kernel) [21]. In this paper, we use 7 COS terms to approximate the PER kernel. The RBF kernel can be approximated by a SS model based on the Matern $d/2$ kernel, where $d = 1, 3, 5, 7, 9, \ldots$ and the approximation improves as d increases. In this paper, we will use $d = 3$.

2.4   Time Series Forecasting and Priors

In [3], GP regression was proposed for time series forecasting using the following composite kernel:

$$K = PER + LIN + RBF + SM_1 + SM_2 + WN. \qquad (9)$$

The periodic kernel (PER) captures the seasonality of the time series. LIN captures the linear trend. Long-term trends are generally smooth, and can be properly modelled by the RBF kernel. The two SM kernels are used to pick up the remaining signal. Finally, the WN kernel represents the observation (Gaussian) noise.

This results in a kernel capturing a wide range of patterns but comprising 16 hyperparameters, which must be estimated from data. This might be challenging on short time series, such as monthly or quarterly ones. In [3] the problem is addressed by setting priors on the hyperparameters. In particular, lognormal priors are adopted and they are defined through a hierarchical Bayes approach, i.e., by analyzing a subset of monthly time series from the M3 competition. The priors, which we also adopt, are given in Table 2.

Table 2. Parameters of the lognormal priors. The same prior is adopted for the variances of all components in Eq. (9)

Parameter	ν	λ
Variance	−1.5	1.0
Lengthscales		
std_periodic	0.2	1.0
rbf	1.1	1.0
SM_1	−0.7	1.0
SM_2	1.1	1.0

2.5 SS Approximation

To achieve $O(n)$ complexity, we replace the kernel in (9) with this approximation

$$\tilde{K} = (+_8 \text{COS}) + \text{LIN} + \text{MAT32} + \text{COS} \times \text{MAT32} + \text{COS} \times \text{MAT32} + \text{WN}. \tag{10}$$

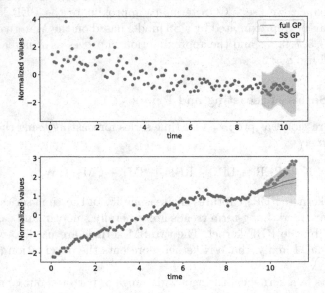

Fig. 1. Comparison of GP and SS forecasts. The blue dots are the training data and the purple dots the test data. The small differences between full GP and SS are due to the slightly different estimation of the hyperparameters. The time series are monthly and the forecasts are computed up to 1.5 years ahead; time is expressed in years.

Note we have approximated PER with the sum of 7 COS kernel and RBF with MAT32.[6] A GP with the above kernel can equivalently be represented by a SS model who state has dimension $7 \times 2 + 2 + 2 + 4 + 4 + 1 = 24$.

Figure 1 compares the GP estimate and forecast based on the kernel (9) and the SS approximation based on the kernel (10) on some time series from the M3 competition.[7] The SS approximation provides close forecasts to the full GP. We provide a more in-depth analysis when discussing the experiments.

2.6 Combining GP Kernel with Exponential Smoothing

Our framework is so flexible, that it allows combining the state-space representations of covariance functions and existing state-space models, thus obtaining some novel time series models.

As a proof of concept, we consider state-space additive exponential smoothing (*additive ets*), and we replace its seasonal component with the PER kernel.

The discrete-time SS representation of exponential smoothing with linear trend is [8]:

$$\text{Holt:} \begin{cases} f_1((k+1)\Delta_t) = f_1(k\Delta_t) + f_2(k\Delta_t) + \alpha w((k+1)\Delta_t) \\ f_2((k+1)\Delta_t) = f_2(k\Delta_t) + \alpha\beta w((k+1)\Delta_t) \\ y((k+1)\Delta_t) = f_1(k\Delta_t) + f_2(k\Delta_t) + w((k+1)\Delta_t) \end{cases} \quad \begin{bmatrix} f_1(t_0) \\ f_2(t_0) \end{bmatrix} \sim \mathcal{N}\left(\begin{bmatrix} 0 \\ 0 \end{bmatrix}, \begin{bmatrix} s_l^2 & 0 \\ 0 & s_b^2 \end{bmatrix} \right)$$

where Δ_t is the sampling frequency and w are independent Gaussian noises with zero mean and variance s_v^2. Such model has five parameters: $\alpha, \beta \in [0,1]$ and s_l^2, s_b^2, s_v^2.

We then complete the SS model by adding the (approximated) SS representation of the PER kernel, constituted by the sum of seven COS covariance functions. When estimating the hyperparameters, automatic relevance determination (ARD) automatically makes irrelevant the unnecessary component, without the need for a separate model selection step.[8]

3   Experiments

We consider the following GP models:

- full-GP: the model of Eq. (9), trained with priors [3];
- full-GP$_0$: the same model, trained by maximizing the marginal likelihood (no priors);

---

[6] We also tried a more accurate approximation of the periodic kernel, 11 COS kernels, but it did not provide a significant better performance in the M3 competition.

[7] In both cases, we have estimated the kernels hyperparameters using MAP.

[8] For the variances of the Holt's model we use the same priors as in Table 2. For α, β, we use the prior Beta(1, 1.4) and, respectively, Beta(1, 11.4). We learned the parameters of these priors using a hierarchical model similar to the one described in [3].

– SS-GP and SS-GP$_0$, i.e., the corresponding SS models (Eq. 10) trained with and without priors.

We use a single restart when training all the models.

As benchmarks, we consider *auto.arima* and *ets*, both available from the forecast package [9]. The *auto.arima* algorithm first makes the time series stationary via differentiation; then it fits an ARMA model selecting the orders via AICc. The *ets* algorithm fits several state-space exponential smoothing models [8], characterized by different types of trend, seasonality and noise; the best model is eventually chosen via AICc. All the considered models represent the forecast uncertainty via a Gaussian distribution.

Metrics. As performance metric, we consider the mean absolute error (MAE) on the test set:

$$\mathrm{MAE} = \sum_{t=1}^{T} |y_t - \hat{y}_t|$$

where we denote by y_t and $\hat{y}_t$ the actual value and the expected value of the time series at time t; σ_t^2 denotes the variance of the forecast at time t and by T the length of the test set.

Furthermore, we compute the continuous-ranked probability score (CRPS) [5], which generalizes the MAE to the case of probabilistic forecasts. It is a proper scoring rule for probabilistic forecasts, which corresponds to the integral of the Brier scores over the continuous predictive distribution. MAE and CRPS are loss functions, hence the lower the better.

3.1 Monthly M3

Table 3. Performance on the M3 monthly time series.

Algorithm	Median		Mean	
	MAE	CRPS	MAE	CRPS
SS-GP	**0.489**	**0.342**	**0.567**	**0.421**
full-GP	**0.482**	**0.347**	**0.565**	**0.414**
SS-GP$_0$	0.550	0.408	0.627	0.499
full-GP$_0$	0.546	0.390	0.628	0.460
ETS	0.516	0.369	0.595	0.436
Auto.arima	0.515	0.373	0.588	0.430

The M3 competition includes 1489 monthly time series. We exclude 350 of them, which were used in [3] to define the priors of Table 2, which we also adopt. We thus run experiments on the remaining 1079 monthly time series. The length of training set varies between 49 and 126 months, while the test set is always 18 months long. We standardize each time series using the mean and the standard deviation of the training set. We fix the period of the periodic kernel to one year, which is standard practice for M3.

The median and mean results for time series are given in Table 3. The SS-GP and full-GP obtain the best median and mean performance on all indicators. The performance of full-GP and of its state-space representation is practically identical, showing that the SS approximation is very accurate. We tried also Prophet [23] but its accuracy was not competitive. We thus dropped it.

The large improvement of full-GP and SS-GP over full-GP$_0$ and SS-GP$_0$ confirms that the priors are necessary to exploit the potential of the GP.

3.2 Combining GP Kernel and Exponential Smoothing

The SS representation of GPs allows us to combine GPs with state-of-the-art models for time series forecasting such the ETS model [8].

In this section, we compare the SS model discussed previously, which uses the following kernel:

$$\tilde{K}_1 = (+_7\text{COS}) + \text{Holt}, \tag{11}$$

where the Holt kernel has been defined in Sect. 2.6.

We compare this model with *additive ETS* model, defined as follows. The additive ets model fits four different models via maximum likelihood and performs model selection via AICc. The four models are simple exponential smoothing (*ses*, no trend and no seasonality), *ses* with linear trend, *ses* with linear trend and additive seasonality, *ses* with additive seasonality but no trend. We implement all such models using the forecast package for R [9]. The *ets* framework makes available also multiplicative models, that however we do not consider in this section.

The seasonal component of exponential smoothing has some shortcomings: it requires to estimate $(m + 1)$ parameters, where m denotes then number of samples within a period (e.g., $m = 12$ for monthly time series); moreover, it does not manage complex seasonalities such non-integer periods or multiple seasonal pattern. In our model we thus substitute it with the PER kernel (equivalently $(+_7\text{COS})$ kernel), which has only two (hyper)-parameters and which can model complex seasonalities (e.g., multiple seasonalities can be modelled by adding multiple PER kernels).

Therefore, the main differences between additive ets and our novel model are thus:

– PER kernel vs seasonal component of exponential smoothing;
– automatic relevance determination vs model selection.

The simulation results are shown in Table 4. SS-GP is again the best model. Comparing SS-GP performance in Table 3 and 4 is evident that the more complex kernel (10) provides a better the performance. However, this shows how the SS

representation of GPs opens up the possibility of developing novel time series
models combining traditional time series models with "machine-learning-like"
models.

Table 4. Performance on M3 monthly. SS-GP with kernel $\tilde{K}_1$ compared to additive
ETS.

	Median		Mean	
Algorithm	MAE	CRPS	MAE	CRPS
SS-GP	**0.511**	**0.368**	**0.581**	**0.436**
SS-GP$_0$	0.538	0.387	0.608	0.461
Additive ETS	0.533	0.381	0.601	0.439

3.3   Large Datasets and Multiple Seasonality

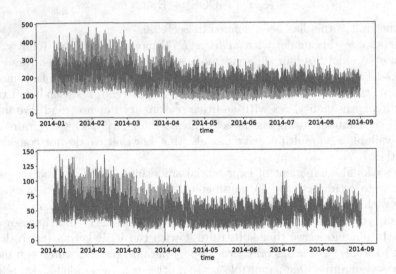

Fig. 2. Two time series taken from the Electricity Dataset

By contrast to full GP, SS models can scale to large datasets. We provide a proof-
of-concept of that by applying the SS model to two time series in the UCI's Electricity Dataset. Each time series is relative to the electricity consumption of client
from a period of 2011 to 2014 at an interval of 15 min. The goal is to forecast the
electricity consumption one week ahead. The length of each time series is 23997
and, therefore, we cannot run full GP (on a standard PC). Moreover, the time

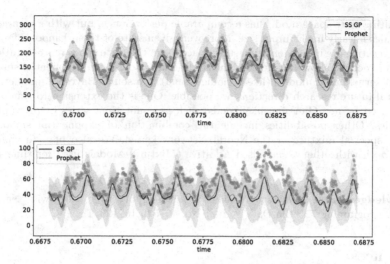

Fig. 3. One week ahead forecast computed by (i) the proposed SS model; (ii) Facebook's Prophet; for the two time series in Fig. 2. The time has been normalized: 1 is one year.

series have both daily and weekly periodicity, which means the kernel in (10) is not appropriate.

However, we can easily deal with multiple seasonality by adding another periodic component to the kernel:

$$\tilde{K} = (+_7\text{COS}) + (+_7\text{COS}) + \text{LIN} + \text{MAT32} + \text{COS} \times \text{MAT32} + \text{COS} \times \text{MAT32} + \text{WN} \tag{12}$$

where the first periodic kernel (the term $(+_7\text{COS})$) has period $1/365.25$ and the second $7/365.25$.[9]

Figure 2 shows two time series taken from the Electricity Dataset. Figure 3 reports the relative one week ahead forecast computed by (i) the proposed SS model; (ii) Facebook's Prophet. The training times are of few seconds for Prophet, and about 300 s for the SS model.

While our implementation is currently slower than Prophet, it already handles flawlessly this time series. The training time of our implementation can be largely reduced by using Stochastic Gradient (SGD) optimization, thus working with minibatch of data. The forecasts show that the SS model is competitive also on long time series; however, the analysis of a large number of time series is needed in order to achieve conclusions which are significant. We defer this analysis to future work, after the completion of a faster implementation of SS-GP based on SGD.

4   Conclusions

Focusing on time series forecasting, we have shown that a Gaussian Process with a complex composite kernel can be accurately approximated by a state space model.

---

[9] By contrast to arima and ETS, GP and SS models can easily model non-integer seasonality like the ones in the Electricity dataset, see [3] for more details.

The resulting state space model has a comparable performance, but with a complexity which scales linearly in the input size. Moreover, given state-of-the-art models for time series forecasting are implemented in state space form, the state space representation of Gaussian Processes allowed us to combine traditional models (like exponential smoothing) with kernel-based models (like periodic kernel) in a sound Bayesian manner.

Several future research directions are possible. One is the extension to time series characterized by non-Gaussian likelihoods, such as count time series or intermittent time series. Other possibilities include the combination of exponential smoothing with the spectral mixture or the Neural Network kernel. We also plan to compare our approach with other Generalised Additive (Mixture) Models used for time-series forecasting.

Acknowledgements. The authors acknowledge support from the Swiss National Research Programme 75 "Big Data" Grant No. 407540_167199/1.

References

1. Bauer, M., van der Wilk, M., Rasmussen, C.E.: Understanding probabilistic sparse Gaussian process approximations. In: Advances in Neural Information Processing Systems, pp. 1533–1541 (2016)
2. Benavoli, A., Zaffalon, M.: State Space representation of non-stationary Gaussian processes. arXiv preprint arXiv:1601.01544 (2016)
3. Corani, G., Benavoli, A., Zaffalon, M.: Time series forecasting with Gaussian Processes needs priors. In: Proceedings of the ECML PKDD (2021, accepted). https://arxiv.org/abs/2009.08102
4. Foreman-Mackey, D., Agol, E., Ambikasaran, S., Angus, R.: Fast and scalable Gaussian process modeling with applications to astronomical time series. Astron. J. **154**(6), 220 (2017)
5. Gneiting, T., Raftery, A.E.: Strictly proper scoring rules, prediction, and estimation. J. Am. Stat. Assoc. **102**(477), 359–378 (2007)
6. Hensman, J., Fusi, N., Lawrence, N.D.: Gaussian processes for big data. In: Proceedings of the Twenty-Ninth Conference on Uncertainty in Artificial Intelligence, UAI 2013, pp. 282–290. AUAI Press, Arlington (2013)
7. Hernández-Lobato, D., Hernández-Lobato, J.M.: Scalable Gaussian process classification via expectation propagation. In: Artificial Intelligence and Statistics, pp. 168–176 (2016)
8. Hyndman, R.J., Athanasopoulos, G.: Forecasting: Principles and Practice, 2nd edn. OTexts, Melbourne (2018). OTexts.com/fpp2
9. Hyndman, R.J., Khandakar, Y.: Automatic time series forecasting: the forecast package for R. J. Stat. Softw. **26**(3), 1–22 (2008). http://www.jstatsoft.org/article/view/v027i03
10. Jazwinski, A.H.: Stochastic Processes and Filtering Theory. Courier Corporation, New York (2007)
11. Karvonen, T., Sarkkä, S.: Approximate state-space Gaussian processes via spectral transformation. In: 2016 IEEE 26th International Workshop on Machine Learning for Signal Processing (MLSP), pp. 1–6. IEEE (2016)
12. Lloyd, J.R.: GEFCom2012 hierarchical load forecasting: gradient boosting machines and Gaussian processes. Int. J. Forecast. **30**(2), 369–374 (2014)

13. Loper, J., Blei, D., Cunningham, J.P., Paninski, L.: General linear-time inference for Gaussian processes on one dimension. arXiv preprint arXiv:2003.05554 (2020)
14. Quiñonero-Candela, J., Rasmussen, C.E.: A unifying view of sparse approximate Gaussian process regression. J. Machine Learn. Res. **6**, 1939–1959 (2005)
15. Rasmussen, C., Williams, C.: Gaussian Processes for Machine Learning. The MIT Press, Cambridge (2006)
16. Roberts, S., Osborne, M., Ebden, M., Reece, S., Gibson, N., Aigrain, S.: Gaussian processes for time-series modelling. Philos. Trans. Royal Soc. A Math. Phys. Eng. Sci. **371**(1984), 20110550 (2013)
17. Särkkä, S., Hartikainen, J.: Infinite-dimensional Kalman filtering approach to spatio-temporal Gaussian process regression. In: International Conference on Artificial Intelligence and Statistics, pp. 993–1001 (2012)
18. Sarkka, S., Solin, A., Hartikainen, J.: Spatiotemporal learning via infinite-dimensional Bayesian filtering and smoothing: a look at Gaussian process regression through kalman filtering. Signal Process. Mag. IEEE **30**(4), 51–61 (2013)
19. Schuerch, M., Azzimonti, D., Benavoli, A., Zaffalon, M.: Recursive estimation for sparse Gaussian process regression. Automatica **120**, 109–127 (2020)
20. Snelson, E., Ghahramani, Z.: Sparse Gaussian processes using pseudo-inputs. In: Advances in Neural Information Processing Systems, pp. 1257–1264 (2006)
21. Solin, A., Särkkä, S.: Explicit link between periodic covariance functions and state space models. In: Artificial Intelligence and Statistics, pp. 904–912. PMLR (2014)
22. Solin, A., Sarkka, S.: Gaussian quadratures for state space approximation of scale mixtures of squared exponential covariance functions. In: 2014 IEEE International Workshop on Machine Learning for Signal Processing (MLSP), pp. 1–6. IEEE (2014)
23. Taylor, S.J., Letham, B.: Forecasting at scale. Am. Stat. **72**(1), 37–45 (2018)
24. Titsias, M.: Variational learning of inducing variables in sparse Gaussian processes. In: van Dyk, D., Welling, M. (eds.) Proceedings of the Twelth International Conference on Artificial Intelligence and Statistics. Proceedings of Machine Learning Research, PMLR, Hilton Clearwater Beach Resort, Clearwater Beach, Florida USA, 16–18 April 2009, vol. 5, pp. 567–574 (2009)
25. Wilson, A., Adams, R.: Gaussian process Kernels for pattern discovery and extrapolation. In: International Conference on Machine Learning, pp. 1067–1075. PMLR (2013)
26. Wood, S.N.: Generalized Additive Models: An Introduction with R. CRC Press, Boca Raton (2017)

Fast Channel Selection for Scalable Multivariate Time Series Classification

Bhaskar Dhariyal[✉], Thach Le Nguyen, and Georgiana Ifrim

School of Computer Science, University College Dublin, Dublin, Ireland
{bhaskar.dhariyal,thach.lenguyen,georgiana.ifrim}@insight-centre.org

Abstract. Multivariate time series record sequences of values using multiple sensors or channels. In the classification task, we have a class label associated with each multivariate time series. For example, a smartwatch captures the activity of a person over time, and there are typically multiple sensors capturing aspects of motion such as acceleration, orientation, heart beat. Existing Multivariate Time Series Classification (MTSC) algorithms do not scale well with large datasets, and this leads to extensive training and prediction times. This problem is attributed to an increase in the number of records (e.g., study participants), duration of recording (time series length), and number of channels (e.g., sensors). Existing MTSC methods do not scale well with the number of channels, and only a few methods can complete their training on the medium sized UEA MTSC benchmark within 7 days. Additionally, for some problems, only a few channels are relevant for the learning task, and thus identifying the relevant channels before training may help with improving both the scalability and accuracy of the classifiers, as well as result in savings for data collection and storage. In this work, we investigate a few channel selection strategies for MTSC and propose a new approach for fast supervised channel selection. The key idea is to use channel-wise class separation estimation using fast computation on centroid-pairs. We evaluate the impact of our new method on the accuracy and scalability of a few state-of-the-art MTSC algorithms and show that our approach can dramatically reduce the input data size, and thus improve scalability, while also preserving accuracy. In some cases, the runtime for training the classifier was reduced to one third of the runtime on the original dataset. We also analyse the performance of our channel selection method in a case study on a human motion classification task and show that we can achieve the same accuracy using only one third of the data.

Keywords: Channel selection · Dimension reduction · Time series

1 Introduction

Time series are data recorded as ordered sequences of numeric values and are encountered in many applications. The proliferation of IoT and sensor technology has rapidly fuelled the collection of such sequential data. Furthermore, the

© Springer Nature Switzerland AG 2021
V. Lemaire et al. (Eds.): AALTD 2021, LNAI 13114, pp. 36–54, 2021.
https://doi.org/10.1007/978-3-030-91445-5_3

onset of the covid19 pandemic has also enhanced the growth of temporal data collection. For instance, a study [1] in March 2021 reported 28% growth in the market for wearable sensing devices. Besides sensors, multimedia files like images, audio, and video can also be converted to time series to save on data storage, and thus become significant contributors to time series database growth.

Time-series applications vary across domains, e.g. sports science, agriculture, or healthcare. For example, a person lifts a barbell above their head from shoulder level in the Military Press exercise. The motion of various body parts during the exercise can be captured to analyse the correctness of the exercise execution. Sensors or video recordings can help track the movement of body parts in the form of temporal data (time series) [18]. The body parts that act as data sources are known as channels in the time series context, and these channels record time-series data simultaneously. The execution of the exercise can be classified into normal and aberrant subtypes. The task of assigning discrete labels to multi-channel time series is known as Multivariate Time Series Classification (MTSC). In the case of a single channel, the task of assigning a label is known as Univariate Time Series Classification (UTSC). Figure 1 illustrates the video capture of a person doing a Military Press exercise and the extraction of multivariate time series using body pose estimation with OpenPose [3].

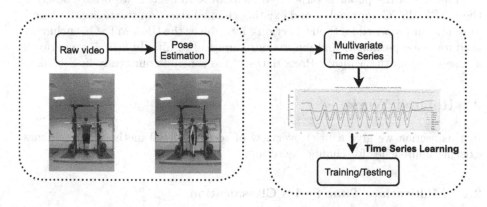

Fig. 1. From video to multivariate time series using OpenPose (figure from [18]).

Research in UTSC has made significant progress [2], but there is much less work done on MTSC [13]. Most literature in UTSC considers the MTSC problem as an extended version of UTSC and tends to adapt UTSC methods for MTSC. However, such methods ignore the computational components such as space and time complexity which are crucial elements for MTSC, thus rendering most of the state-of-the-art (SOTA) classifiers infeasible for practical applications. The recent studies [5,13] highlighted scalability as a big challenge for SOTA classifiers in MTSC, with many existing algorithms not able to complete training on 26 medium-sized UCR MTSC datasets within 7 days. The scalability challenge can be analysed from three perspectives: the number of channels, length of time series and number of samples in the dataset. This study focuses on the first aspect and proposes a new method to select relevant channels from the training data, before training a classifier. The study's primary objective is to enable existing SOTA classifiers to scale better with an increase in the number of MTSC channels, by reducing the time and memory required for computation, while maintaining accuracy. In particular, we examine the impact of our channel selection approach on the recent MTSC algorithms Rocket [4], MrSEQL [10], Weasel-MUSE [16] and 1NN-DTW [13]. The main contributions of this study are:

- We propose three greedy channel selection strategies for MTSC, to scale up existing MTSC algorithms.
- We conduct extensive experiments on the UCR MTSC benchmark and report a 70% reduction in computation time for the combination of channel selection plus training MTSC algorithms, while preserving the classifier accuracy.
- We show that not all the data is useful for classification and that we achieve significant data storage savings, e.g., 70% of the original data can be discarded with our approaches.
- We present a case study of our methods on a real-world, 25-channel MTSC dataset, recorded for the Military Press strength and conditioning exercise.

The rest of the paper is structured as follows. In Sect. 2, we briefly describe the SOTA MSTC approaches and existing channel selection strategies. Section 3 presents our proposed methods. Section 4 introduces the UEA MTSC benchmark used for the experiments and reports our empirical results. In Sect. 5 we perform a case study on the Military Press dataset. We conclude our study in Sect. 6.

2   Related Work

In this section we give a brief overview of recent MTSC methods and discuss existing approaches for channel selection.

2.1   Multivariate Time Series Classification

The recent empirical survey [13] provides a detailed overview of progress in MTSC. Here we describe a subset of those methods, with a specific focus on methods that were shown to complete the training and testing on the 26 equal-length UEA MTSC datasets within 7 days and do not require advanced HPC infrastructure.

1NN-DTW is a 1-Nearest Neighbour classifier with Dynamic Time Warping (DTW) distance and one of the most popular methods in MTSC. In [17] the authors proposed two versions of DTW for MTS data, DTW_I and DTW_D, to study the impact of DTW on multiple channels. The main difference between the two versions is how they compute the distance between two multivariate time series. The DTW_I assumes each channel of MTS as an independent univariate time-series and consequently sums up the distances for each channel pair.

It calculates the optimal path P based on the pointwise distance between the time series. The DTW_D assumes that the correct warping is the same across all channels; it computes the distance between two time series by first summing up the distance across each channel. Unlike for DTW_I, in DTW_D the optimal path P is based on the euclidean distance between two vectors that represent all channels. Although both versions of DTW have high accuracy and are considered a strong baseline for any MTSC task, they are computationally expensive[1] and have been outperformed in accuracy by more recent methods [13]. Both DTW_I and DTW_D are heavily impacted by the number of time series channels, thus optimising the number of channels can drastically help in improving their scalability.

MrSEQL-SAX [10] is a linear classifier that extracts symbolic features from time series. The method first transforms time-series data to multiple symbolic representations of different domains (e.g., SAX [11] in the time domain and SFA [15] in the frequency domain) and different resolutions (i.e., different window sizes). The classifier extracts discriminative subsequences from the symbolic representations and these subsequences are later combined to form a feature vector used to train a classification model. In [10] the authors showed that adding features from different representations types (e.g., SAX and SFA) boosts the accuracy of the classifier. The method was initially developed for UTSC but was also adapted for MTSC; the adapted version views each channel as an independent representation of the time series. Unlike UTSC ensemble classifiers, MrSEQL uses all the extracted features from the different representations by combining them into a single feature space to train the final model. The SAX/SFA symbolic transforms are computationally expensive. Furthermore, transformation over multiple windows iterating over full time series incurs a high cost to the scalability of the classifier. Therefore, reducing the number of channels has potential to significantly improve the scalability of MrSEQL. MrSEQL-SAX is a version of MrSEQL restricted to use only SAX features (this version is more efficient than the one using both SAX and SFA features).

WEASEL-MUSE [16] is an extension of the WEASEL algorithm developed for UTSC. The classifier builds a bag-of-pattern (BOP) model using the SFA transform for every channel. By rolling multiple windows of varying size on raw and derivative time series, this method transforms those segments into unigram and bigram words. The classifier links these words to their respective channel and creates a histogram for each channel separately. Since there are many features for every channel, the Chi-square feature selection method removes the irrelevant ones. The selected features are concatenated into a single feature vector which is fed to a logistic regression algorithm. Similar to MrSEQL, the WEASEL-MUSE classifier also iterates over the entire time series for every channel and performs the SFA transform for every window. This iteration and transformation

---

[1] We have evaluated here the sktime implementation of DTW.

increase the overall computation cost. Also, storing many unigrams and bigrams in memory is quite expensive.

ROCKET [4] is a recent classifier initially developed for UTSC and also extended to MTSC. Models produced by ROCKET are often highly accurate while keeping the computational burden low. Inspired by CNN, ROCKET relies on convolutional kernels to extract features. Instead of learning the weights of kernels through backpropagation, ROCKET randomly generates many convolutional kernels. Additionally, kernel properties like length, dilation, stride, bias and zero-padding are sampled at random. For the MTSC task, the internal channel selection is also random. A close look at the code shows that there are at most 12 channels selected for any MTSC dataset, so the runtime of this algorithm is not significantly affected by the number of channels, especially for datasets with more than 12 channels. In principle, the kernels are just a simple linear transformation of the input time series producing a new time series. A linear classifier, ridge regression, trains on the feature vector formed by global max pooling and the proportion of positive values (PPV) features extracted from the convolution time series from every channel. ROCKET has become very popular due to its high accuracy and speed, yet the impact of the number of channels on this classifier in the MTSC task is not yet examined.

2.2   Channel Selection for Multivariate Time Series Classification

Channel Selection for multivariate time series is a recurring topic in the MTSC literature. However, the focus of most work has been on accuracy, rather than scalability. The most recent work on channel selection [8] tries to identify the best subset of channels. The author's method calculates a merit score based on correlation patterns of the outputs from the classifiers. The algorithm iterates through every possible subset to calculate the merit score, followed by wrapper search on the subset with top 5% merit score. 1NN-DTW is employed to perform the classification. DTW is computationally expensive [5], and using DTW over every possible subset amplifies this problem. Another notable study is CleVer [19] where the author proposed three unsupervised feature subset selection techniques employing Common Principal Component Analysis (CPCA) [9] to measure the importance of each sensor. The authors build a correlation coefficient matrix among different channels for each MTS. The principal components of each coefficient matrix are calculated, and all the principal components are aggregated together, and descriptive component principal components are calculated. The l_2-norm of the resulting vector generates the rank of each channel. The work [7] proposed a framework for channel selection using a voting-based method. The two criteria used were distance-based classification and confidence-based classification. These methods were proposed for streaming data, which is outside the scope of the current study.

In the recent study [6], the authors presented an algorithm for channel ranking and channel selection. The key idea is that if a channel produces similar time series with the same label and different time series with a different label, then it is an informative channel. The channel ranking algorithm assigns a relevance score for each channel. The relevance scores are constructed on a similarity graph among the channels. The authors find the largest eigenvector of the normalized adjacency matrix of the similarity graph, which reflects its cluster structure. Apart from channel ranking the authors also propose channel subset selection. From the adjacency matrix above, the algorithm finds the linear combination of matrices that approximates the similarity matrix of the labels and use the minimum number of redundant channels. Although the proposed channel ranking and selection approach performs well with regards to accuracy, it is slow. Some of the computational bottlenecks include finding the eigenvector and using DTW as the distance measure.

3   Proposed Methods

Let $X \in R^{n \times d \times l}$ be an MTS dataset and y the labels of the time series in the dataset. We denote by n the number of time series in the dataset, d the number of channels in the multivariate time series and l the time series length. In this paper, we only consider fixed-length time series datasets.

The proposed channel selection method makes use of class centroids as representatives of the classes. Let $X_A = \{t \in X \mid y(t) == A\}$ be the subset of X that contains only samples from class A. The centroid of class A is computed as the average of time series in that class:

$$C_A[i,j] = \frac{\sum_{k=1}^{k=m} X_A[k,i,j]}{m}$$

where m is the number of samples in class A. The multi-channel centroid C_A is a $d \times l$ matrix in which each row $C_{A,i}$ is the centroid of class A for channel i.

The centroid-driven channel selection technique computes the distance matrix for every pair of class centroids, for each channel, using a distance function(Δ) discussed later in the section. For a dataset with r classes, the total number of pairs of class centroids is $\frac{r*(r-1)}{2}$. For instance, the distance matrix for a 4-channel dataset with four classes is shown in Table 1, where channel $RElbow$ has the highest distance (166.99) for the pair of class centroids C_n and C_r. In our proposed method, we examine this distance matrix to select the channels that are most likely to be useful. The idea is that channels with larger distances between class centroids are more likely to be discriminative, since centroids behave like prototypes for time series in those classes. In this example, the distance between class centroids C_n and C_r is highest for channel $RElbow$ therefore this channel is more likely to be useful in separating these classes than the other channels, while channel $Nose$ has small distances for all class pairs and so does not seem to be useful to separate any classes. Our method has three components: a distance measure used to compare centroids, an elbow cut heuristic used to threshold the ranked list of channels and three channel selection strategies.

Table 1. Illustration of a distance matrix with 4 channels and 4 classes: $a, arch, n, r$. More details about this dataset are provided in the case study described in Sect. 5.

Channels	$\Delta(C_a, C_{arch})$	$\Delta(C_a, C_n)$	$\Delta(C_a, C_r)$	$\Delta(C_{arch}, C_n)$	$\Delta(C_{arch}, C_r)$	$\Delta(C_n, C_r)$
Nose	15.93	11.13	14.90	14.20	14.60	16.93
RElbow	34.62	48.95	157.12	33.04	153.80	166.99
RHeel	29.66	39.84	9.15	16.65	25.86	35.88
LWrist	38.557	42.95	148.48	47.40	155.56	157.55

Distance Metric. In our current work we use euclidean distance to calculate the distance (Δ) between the centroid pairs for each channel. The Euclidean distance is measured as the $l2$ norm of the difference between the centroids.

$$\Delta(C_{A,i}, C_{B,i}) = \|C_{A,i} - C_{B,i}\|$$

Channel Selection Strategies. We propose and evaluate three different strategies for channel selection to identify useful channels.

- **KMeans.** This strategy applies k-means clustering with $k = 2$ on the distance matrix to segregate the channels. Every channel (row from distance matrix) is assigned to one cluster. The cluster centroid represents the mean distance of channels across every class pair. Thus, the mean of the cluster centroid acts as a discriminating criterion. We select the channels from the cluster whose centroid-mean is greater than the other centroid-mean, meaning that this cluster contains channels with higher separation distance, while the other cluster contains noisy channels.
- **Elbow Class Sum (ECS).** From the distance matrix, we sum all the pairwise distances for each channel (sum each row). The sum of the distances is sorted in descending order, and an elbow-point is retrieved using the elbow-cut approach described below. All the channels with a distance higher than that of the elbow point are selected as the relevant channels. A single large centroid-pair distances can bias this type of channel selection, favouring channels that separate two classes clearly, but may not be useful for separating other classes.
- **Elbow Class Pairwise (ECP).** The second strategy, ECS, can be biased towards channels that are useful for separating only a few classes. An alternative strategy iterates through every class pair, selects the best set of channels for that pair and finally takes the union of channels over all pairs. This eliminates the potential bias found in the previous strategy. In some cases, this can lead to selecting all the channels, however, there were only few instances of this behaviour in the UEA dataset.

Elbow Cut. The elbow cut method [14] is a method to determine a point in a curve where significant change can be observed, e.g., from a steep slope to almost flat curve. This point is often referred to as the elbow or the knee point. This is a well-known method to determine the best number of clusters when doing clustering. We apply it here to separate useful channels from noisy channels. An algorithm takes as input the sorted distances corresponding to channels and returns the elbow point. The elbow is the point at the highest distance(d) from the line(b) joining the initial and ending point as shown in Fig. 2. The distance d to any point on the distance curve is calculated as $d = |p - (p.\hat{b})\hat{b}|$ where $\hat{b} = \frac{b}{\|b\|}$ and $p.\hat{b}$ is the projection of p onto $\hat{b}$. The elbow-point is then elbow $= argmax(d)$. The channels that come before the *elbow* are selected as useful channels for classification and the smaller dataset with this subset of channels is used for the classification step (see Algorithm 1). It is clear from Fig. 2 that the elbow point can be relaxed thus allowing a trade-off between data storage and the accuracy of classification. In our work we use the first elbow point which corresponds to channel RShoulder and select only the channels before this point.

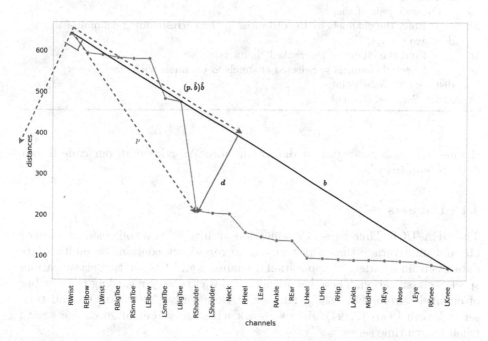

Fig. 2. Elbow-point channel selection. All the channels up to RShoulder are selected.

4   Evaluation

All the experiments were conducted using the popular Python library sktime [12]. Our primary objective in designing experiments is to understand the relative gain or loss in computational aspects of MTSC algorithms using the proposed

---

Algorithm 1: Channel Selection for an MTSC dataset.

Input: Train dataset: X,y
Output: Selected channels
1 Initialization;
2 For each channel in X and each class, compute class centroid ;
3 Compute distance matrix for all pair of centroids;
4 if Channel-Selection is KMeans;
5   Create 2 clusters using KMeans;
6   Selected channels = cluster with higher centroid mean;
7 elif Channel-Selection is ECS ;
8   Sum the distance matrix by rows;
9   Rank channels by sum distance;
10   Find the elbow on the ranked channels;
11   Selected channels = channels with sum distances > elbow point ;
12 elif Channel-Selection is ECP;
13   For each pair of classes;
14   Rank the channels by the distances in the corresponding column of the distance matrix;
15   Find the elbow on the ranked channels;
16   Selected channels = Selected channels ∪ channels with class pair-wise distance > elbow point;
17 Return Selected channels;

---

channel selection strategies for data reduction. We release all our code in our Github repository[2].

4.1 Datasets

The UEA/UCR Time Series Classification archive [2] is a collection of univariate and multivariate time series data. The repository contains 30 multivariate datasets from a variety of application domains, e.g., ECG, motion classification, spectra classification. These heterogeneous datasets vary regarding the number of channels (from 2 to 1,345), number of time series (12 to 30,000) and time series length (8 to 17,894). Here we work with the subset of 26 datasets with equal-length time series.

4.2 MTSC Algorithms

All algorithms described in Sect. 2, except ROCKET, utilise all the channels from the MTSC datasets. Table 2 gives the hyperparameter settings used for all the classifiers in this study.

Table 3 presents the results of these MTSC algorithms on the 26 datasets when no explicit channel selection is implemented. The time is shown in hours and is the total time taken by the algorithm for training and prediction. We

---

[2] https://github.com/mlgig/Channel-Selection-MTSC.

Table 2. Hyperparamter setting used for various SOTA methods

Classifiers	Hyperparameter-setting
WEASEL-MUSE	MUSE(random_state=0)
MrSEQL-SAX	MrSEQLClassifier(seql_mode=fs, symrep= ['sax'])
ROCKET	Rocket(random_state=0)
ROCKET*	ROCKET with all channels
1NN-DTW	KNeighborsTimeSeriesClassifier(n_neighbors=1, distance="dtw")

observe that ROCKET and WEASEL-MUSE have almost similar mean accuracy, however ROCKET is much faster. ROCKET implements a random channel selection strategy which allows it to keep the runtime bounded, no matter how many channels the dataset has; we discuss this in more detail later in this section. ROCKET also uses a multi-threaded implementation, while all the other algorithms are single-thread implementations, hence the significant difference in runtime. The baseline 1NN-DTW is the least accurate and the slowest method among the four. MrSEQL-SAX does not use the SFA representations in this study, due to the SFA implementation in sktime being too computationally expensive, so its accuracy is lower than ROCKET and WEASEL-MUSE in this experiment. Both WEASEL-MUSE and MrSEQL are impacted by the runtime taken by the symbolic transform, while 1NN-DTW is impacted by the DTW computation. In the Appendix we provide detailed results with each algorithm on each dataset.

Table 3. Mean accuracy and total time of SOTA on 26 UEA MTSC datasets.

Classifier	Accuracy	Time (in hrs)
ROCKET	71.59	0.1
WEASEL-MUSE	70.28	73.22
MrSEQL-SAX	66.99	141.40
1NN-DTW	65.38	152.07

4.3   MTSC with Channel Selection

Our proposed channel selection strategies (KMeans, ECS, and ECP) are evaluated on the same 26 datasets as above. The channel selection algorithm is run before the MTSC algorithm and it typically results in a reduced dataset for training/testing. We investigate how these strategies impact the classification accuracy, running time (training and testing) and data storage size.

Ratio of Channels Selected. Figure 3 reports the ratio of channels selected by our methods for each dataset (1.0 means no channel is discarded). The acronyms

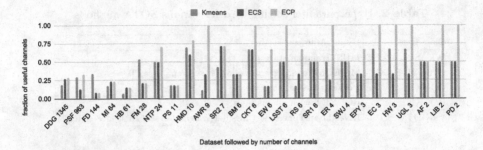

Fig. 3. Fraction of channels selected by each of three channel selection strategies.

and details of the datasets can be found in the Appendix. The ECP ratios appear to be higher in general, as expected and mentioned in Sect. 3. However, this mostly occurs in datasets with a small number of channels (the right side of Fig. 3). On the left side, where the large numbers of channels can become an issue, ECP appears to be just as efficient as the other methods. We also observe that all three methods are more effective on datasets with a larger number of channels that usually pose a significant scalability challenge to existing MTSC algorithms.

In Table 4 we show the total time taken by the three channel selection strategies. This includes the time taken by each method to compute centroids and create the distance matrix. All three techniques are run only on the training dataset and the output is a selected subset of channels. Since these methods only require the distance matrix for centroid-pairs, they are extremely fast even for large datasets, as the time complexity is only affected by the number of classes, and not by the number of samples. The subset of selected channels is then used to create a reduced dataset as input to MTSC algorithms.

Table 4. Total time taken by three channel selection strategies on 26 UEA datasets.

Channel selection strategy	KMeans	ECS	ECP
Total time (minutes)	0.34	0.33	0.35

Performance of Channel Selection. Table 5 shows the change in accuracy and the percentage of time saved by the MTSC algorithms when run on the reduced datasets after applying the three channels selection strategies.

The comparison reveals that there is a massive gain in computation time for a minimal drop in accuracy. The time taken to find the subset (Table 4) is insignificant in comparison. Out of the three channels selection strategies, ECP seems to be the best choice for channel selection. It significantly reduces the computation time and at the same time eliminates noisy channels, thus increasing the

Table 5. Loss/Gain in mean accuracy (ΔAcc) vs percentage time saved (%Time) with respect to All channels (Table 3) for our three channel selection techniques on 26 UCR datasets. The red and blue color indicates loss and gain in accuracy respectively. Higher value for %Time or %Storage indicates more time or storage saved.

Channel Selection→	KMeans	ECS	ECP
Classifiers↓	ΔAcc | %Time	ΔAcc|%Time	ΔAcc | %Time
ROCKET	-4.01 | 33.62	-4.40 | 29.23	+0.13 | 21.43
WEASEL-MUSE	-4.63 | 70.46	-3.80 | 79.90	-1.53 | 73.21
MrSEQL-SAX	-3.33 | 72.68	-3.80 | 84.00	+0.45 | 77.06
1NN-DTW	-4.28 | 68.30	-6.08 | 68.82	+0.67 | 44.80
Mean ΔAcc| Mean %Time	-4.06 | 61.26	-4.52| 65.48	-0.07 | 54.12
Mean %Storage Saved	73.95%	82.59%	74.38%

accuracy for ROCKET, 1NN-DTW and MrSEQL-SAX. The method WEASEL-MUSE takes a small hit on accuracy (1.5%), at the benefit of saving 73.21% in runtime. Considering that WEASEL-MUSE requires 73.2 h to complete training and prediction on this benchmark (see Table 3), this is a significant time saving. A similar result holds across all classifiers, and all channel selection strategies: for a small loss in accuracy, there is a high gain in runtime. In the case of ECP, the accuracy is preserved or even increased, with a significant saving in runtime. We also calculate the average amount of memory saved by the channel selection techniques over the 26 datasets. The comparison of dataframe size in memory, before and after channel selection is used to compute these values. Overall, this MTSC archive uses about 1.6 Gb memory and when using our channel selection strategies, this is reduced to less than 30% of the original size. When stored on disk this dataset is about 3.3 Gb total, and with the channel selection techniques this is reduced to about 900 Mb.

4.4 Effectiveness of Channel Selection

In this experiment we test whether our best strategy (ECP) selects useful channels and how good the selection is compared to selecting optimal channel subsets.

Optimal Channel Subset Selection. We evaluate every possible subset of channels on the test set to discover the optimal subset. Naturally, this brute-force approach is very expensive and impractical for datasets with a high number of channels as the possible combination for a dataset with d channels will be $2^d - 1$. Nevertheless, in this study, all the subsets for datasets with a number of channels <4 are analysed. These datasets are: *AtrialFibrillation, Libras, PenDigits, EthanolConcentration, Epilepsy, Handwriting, UWaveGestureLibrary*. In this experiment, we choose the state-of-the-art ROCKET classifier to quickly evaluate all the subsets. However, because ROCKET internally randomly samples the channels, it can select a good subset by chance and mask the issue of selecting bad channels. Therefore, we modify its code to get ROCKET to use all channels in each kernel, i.e., we use the ROCKET* variant. By doing so, the impact of a good channel subset and a bad channel subset on classification accuracy becomes more pronounced.

Table 6. Accuracy of ROCKET* on datasets with channels <4. Bold indicates the optimal subset. Underscore indicates the subset selected by ECP. Empty spaces are for datasets with less channels, e.g., dataset AF only has 2 channels, 0 and 1.

DT	0	1	2	(0, 1)	(0, 2)	(1, 2)	(0, 1, 2)
AF	**20**	6.67		13.33			
LB	73.9	77.78		**93.89**			
PD	89.59	88.45		**98.26**			
EC	54.4	49.8	**53.6**	38.0	44.9	39.2	36.1
EP	97.82	**100**	94.93	98.55	98.55	97.83	99.28
HW	38.12	32.35	42.12	45.76	59.76	50.35	**57.06**
UW	79.37	71.25	71.88	87.5	93.12	84.06	**93.75**

Table 6 shows that ECP successfully identified the optimal subset five out of seven times. With the Epilepsy (EP) problem, it also correctly identified channel 0 as a potential issue (the classification accuracy is only 97.82% with channel 0 alone) and excluded it from the selection. However, for this dataset it seems to be better to use either only channel 1 or all the channels. It is important to remind the reader that this setting is evaluated directly on the test data, and in practice we do not have perfect knowledge of the best subset of channels for the test data. ECP selects this channels based on the training data alone, and it seems to be effective at finding the useful channels for each task using only training data.

Random Channel Subset Selection. In order to further understand the effect of the ECP channel selection method, we compare the accuracy of the ROCKET classifier, when using channels selected with different strategies. We compare ECPRocket (ECP combined with ROCKET) with ECPsizeRandom-Rocket, a simple baseline where the number of channels is set using ECP, but the actual channels are picked randomly. We repeated the experiment 10 times for each dataset and report the average accuracy in Fig. 4. We observe that for the majority of the large datasets (number of channels >10), ECPRocket is better, while for datasets with less number of channels (number of channels ≤10) the ECPSizeRandomRocket works similar to ECPRocket. Note that for half of the datasets with number of channels ≤10, ECP does not reduce the number of channels (i.e., it keeps all the channels as shown in Fig. 3), hence the two variants ECPRocket and ECPSizeRandomRocket simply reduce to ROCKET, since ECP has no effect in this case. For datasets with a higher number of channels, ECP often reduces the full channel set to a subset of good channels, and the variant ECPRocket constrains ROCKET to work with this pool of good channels, resulting in storage savings and improvements in accuracy. Hence, for either small or large number of channels, ECP is fast and leads to storage savings without resulting in loss of accuracy.

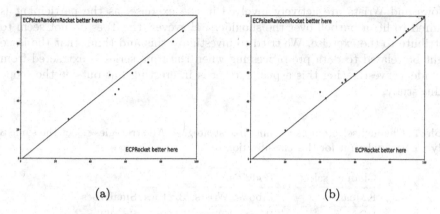

(a)   (b)

Fig. 4. Comparison of ECPRocket with ECPsizeRandomRocket. Figure 4(a) represents datasets with number of channels >10 and Fig. 4(b) represents datasets with number of channels <=10.

5   Case Study: Channel Selection for the Military Press MTSC Dataset

5.1   Dataset

A total of 56 healthy volunteers (34 males and 22 females; age: 26 ± 5 years, height: 1.73 ± 0.09 m and body mass: 72 ± 15 kg) participated in a study aimed at analysing the execution of the Military Press strength and conditioning exercise. The participants completed ten repetitions of the normal form and ten repetitions of induced forms. The NSCA guidelines were applied under the guidance of sports physiotherapists and conditioning coaches to ensure standardisation. The dataset was extracted from the video of individuals performing the exercise with the help of the human body pose estimation OpenPose[3]. There are four classes in the dataset, namely: Normal (N), Asymmetrical (A), Reduced Range (R) and Arch (arch). The N refers to the correct execution of the exercise; A refers to when the barbell is lopsided and asymmetrical, R refers to the form where the bar is not brought down completely to the shoulder level and Arch refers to when participants arch their back. A total of 25 body parts were tracked,

---

[3] https://github.com/CMU-Perceptual-Computing-Lab/openpose.

as seen in Fig. 2. These 25 body parts act as channels for the MTSC task. The train and test size for this dataset is 1452 and 601 respectively and the length of time-series is 160.

5.2   Channel Selection

Table 7 illustrates the selected channels for the Military Press dataset. The Elbows and Wrists are actively involved in the exercise, as the participant is required to lift a barbell over the shoulders. However, the Toes do not seem to contribute to the exercise. We tried to investigate this and think that the issue might be related to data pre-processing when the time series is extracted from the video; investigating this aspect further is interesting but outside the scope of this study.

Table 7. Channel selection using our three strategies. All strategies select the same 8 body parts as relevant for this classification task.

Channel selection	Body parts
KMeans	Elbows, Wrists, BigToes, SmallToes
ECS	Wrists, Elbows, BigToes, SmallToes
ECP	Elbows, Wrists, BigToes, SmallToes

5.3   Results and Discussion

Table 8 reports the results for ECP with different SOTA MTSC classifiers. ROCKET is the fastest and most accurate classifier in this experiment. The data normalisation which is turned on by default in ROCKET, is turned off in the current experiment. This is due to the fact that the signal magnitude contains important information for this task, so normalisation should not be used in this case. For WEASEL-MUSE and MrSEQL-SAX, data normalisation is done internally in the algorithm during the symbolic transform (SFA/SAX), so we cannot de-activate the data normalisation step. This affects the accuracy of these methods in this task, since the magnitude of the signal is important to differentiate between classes. As in the previous experiments, in the case study we also find that ECP saves a large amount of time and memory, with minimal or no loss in accuracy. For WEASEL-MUSE, it saves about 71.6% of computation time, while for MrSEQL-SAX and 1NN-DTW it saved about 74% and 68%, respectively. Moreover, the memory required for computation is reduced to 32%, thus a saving of 68% on the original dataset.

Table 8. Performance of ECP on the Military Press exercise.

Classifiers	Accuracy	Time (minutes)
	ECP | All	ECP | All
ROCKET	76.26 | 77.53	2.14 | 2.25
WEASEL-MUSE	57.57 | 57.57	30.29 | 107.02
MrSEQL-SAX	58.23 | 61.56	139.53 | 516.79
1NN-DTW	48.58 | 47.25	10.39 | 24.36
Data size (MB) Reduced|Original	15.77 | 49.29	

6   Conclusion

In this study we have shown that not all the channels for MTSC are helpful. Data noise in the form of uninformative channels can prevent the classifier from achieving its maximum potential. We have observed that channel selection can remove some of the noise and drastically reduce the required computation time for existing MTSC methods. In the current study, we showed that the distance between the class centroids of various channels plays a crucial role in identifying the noisy channels. Our three-channel selection strategies ECP, ECS and Kmeans, can select the useful channels based on this distance. All three techniques significantly reduced the runtime and memory required to run SOTA classifiers. The ECS and KMeans techniques also reduced the accuracy, while ECP resulted in accuracy gains for MrSEQL-SAX, ROCKET and 1NN-DTW and marginal accuracy loss for WEASEL-MUSE. We believe that with a more robust elbow selection heuristic the performance can be improved further. Our channel selection techniques significantly reduced the data size on disk for most of the MTSC datasets, thus enabling significant storage savings for large MTSC datasets where several channels are not useful for the classification task.

Acknowledgments. This publication has emanated from research supported in part by a grant from Science Foundation Ireland through the VistaMilk SFI Research Centre (SFI/16/RC/3835) and the Insight Centre for Data Analytics (12/RC/2289_P2). For the purpose of Open Access, the author has applied a CC BY public copyright licence to any Author Accepted Manuscript version arising from this submission. We would like to thank the reviewers for their constructive feedback. We would like to thank all the researchers that have contributed open source code and datasets to the UEA MTSC Archive and especially, we want to thank the groups at UEA and UCR who continue to maintain and expand the archive.

Appendix

See Tables 9 and 10.

Table 9. Detailed description for the 26 MTSC datasets used in this study.

Dataset	Acronym	TrainSize	TestSize	NumChannels	SeriesLength	NumClasses	ClassCounts
ArticularyWordRecognition	AWR	275	300	9	144	25	11
AtrialFibrillation	AF	15	15	2	640	3	5
BasicMotions	BM	40	40	6	100	4	10
Cricket	CKT	108	72	6	1197	12	9
DuckDuckGeese	DDG	50	50	1345	270	5	10
EigenWorms	EW	128	131	6	17984	5	55
Epilepsy	EP	137	138	3	206	4	34
ERing	ER	30	270	4	65	6	5
EthanolConcentration	EC	261	263	3	1751	4	65
FaceDetection	FD	5890	3524	144	62	2	2945
FingerMovements	FM	316	100	28	50	2	159
HandMovementDirection	HMD	160	74	10	400	4	40
Handwriting	HW	150	850	3	152	26	8
Heartbeat	HB	204	205	61	405	2	57
Libras	LB	180	180	2	45	15	12
LSST	LSST	2459	2466	6	36	14	34
MotorImagery	MI	278	100	64	3000	2	139
NATOPS	NTP	180	180	24	51	6	30
PEMS-SF	PSF	267	173	963	144	7	32
PenDigits	PD	7494	3498	2	8	10	780
PhonemeSpectra	PS	3315	3353	11	217	39	85
RacketSports	RS	151	152	6	30	4	39
SelfRegulationSCP1	SR1	268	293	6	896	2	135
SelfRegulationSCP2	SR2	200	180	7	1152	2	100
StandWalkJump	SWJ	12	15	4	2500	3	4
UWaveGestureLibrary	UW	120	320	3	315	8	15

Table 10. The amount of memory (MB) used by each dataset when using all channels and after applying our channel selection strategies.

Dataset	OriginalSize	KMeansReduced	ECSReduced	ECPReduced	KMeansSaved%	ECSSaved%	ECPSaved%	Channels
DuckDuckGeese	147.25	28.03	40.40	42.37	80.97	72.56	71.23	1345
PEMS-SF	315.83	92.81	41.00	104.29	70.61	87.02	66.98	963
FaceDetection	511.20	173.95	42.60	42.60	65.97	91.67	91.67	144
MotorImagery	409.53	70.39	95.98	95.98	82.81	76.56	76.56	64
Heartbeat	40.06	2.63	5.91	5.91	93.44	85.25	85.25	61
FingerMovements	4.52	2.42	0.97	0.97	46.43	78.57	78.57	28
NATOPS	2.24	1.12	1.12	1.59	50.00	50.00	29.17	24
PhonemeSpectra	65.10	11.84	11.84	11.84	81.82	81.82	81.82	11
HandMovementDirection	5.09	3.56	3.05	4.07	30.00	40.00	20.00	10
ArticularyWordRecognition	3.04	0.34	1.01	3.04	88.89	66.66	0.00	9
SelfRegulationSCP2	12.49	5.35	8.92	8.92	57.14	28.57	28.57	7
BasicMotions	0.21	0.07	0.07	0.07	66.63	66.63	66.63	6
Cricket	6.00	4.00	4.00	6.00	33.33	33.33	0.00	6
EigenWorms	105.47	17.58	17.58	70.32	83.33	83.33	33.33	6
LSST	5.97	2.98	2.98	5.97	50.00	50.00	0.00	6
RacketSports	0.32	0.05	0.11	0.22	83.30	66.64	33.32	6
SelfRegulationSCP1	11.20	5.60	5.60	5.60	50.00	50.00	50.00	6
ERing	0.08	0.04	0.02	0.08	49.92	74.88	0.00	4
StandWalkJump	0.92	0.46	0.46	0.46	49.99	49.99	49.99	4
Epilepsy	0.70	0.23	0.23	0.47	66.66	66.66	33.33	3
EthanolConcentration	10.56	7.04	3.52	10.56	33.33	66.67	0.00	3
Handwriting	0.58	0.39	0.19	0.58	33.33	66.65	0.00	3
UWaveGestureLibrary	0.91	0.61	0.30	0.91	33.33	66.66	0.00	3
AtrialFibrillation	0.15	0.08	0.08	0.08	49.96	49.96	49.96	2
Libras	0.17	0.09	0.09	0.17	49.96	49.96	0.00	2
PenDigits	2.86	1.43	1.43	2.86	50.00	50.00	0.00	2

References

1. Consumer enthusiasm for wearable devices drives the market to 28.4% growth in 2020 (2021). https://www.idc.com/getdoc.jsp?containerId=prUS47534521
2. Bagnall, A., Lines, J., Bostrom, A., Large, J., Keogh, E.: The great time series classification bake off: a review and experimental evaluation of recent algorithmic advances. Data Min. Knowl. Disc. **31**(3), 606–660 (2016). https://doi.org/10.1007/s10618-016-0483-9
3. Cao, Z., Hidalgo Martinez, G., Simon, T., Wei, S., Sheikh, Y.A.: OpenPose: real-time multi-person 2d pose estimation using part affinity fields. IEEE Trans. Pattern Anal. Mach. Intell. **43**, 172–186 (2019)
4. Dempster, A., Petitjean, F., Webb, G.I.: ROCKET: exceptionally fast and accurate time series classification using random convolutional kernels. Data Min. Knowl. Disc. **34**, 1–42 (2020)
5. Dhariyal, B., Le Nguyen, T., Gsponer, S., Ifrim, G.: An examination of the state-of-the-art for multivariate time series classification. In: 2020 International Conference on Data Mining Workshops (ICDMW), pp. 243–250 (2020). https://doi.org/10.1109/ICDMW51313.2020.00042
6. Han, S., Niculescu-Mizil, A.: Supervised feature subset selection and feature ranking for multivariate time series without feature extraction. arXiv preprint arXiv:2005.00259 (2020)
7. Hu, B., Chen, Y., Zakaria, J., Ulanova, L., Keogh, E.: Classification of multi-dimensional streaming time series by weighting each classifier's track record. In: 2013 IEEE 13th International Conference on Data Mining, pp. 281–290 (2013). https://doi.org/10.1109/ICDM.2013.33
8. Kathirgamanathan, B., Cunningham, P.: A feature selection method for multi-dimension time-series data. In: Lemaire, V., Malinowski, S., Bagnall, A., Guyet, T., Tavenard, R., Ifrim, G. (eds.) AALTD 2020. LNCS (LNAI), vol. 12588, pp. 220–231. Springer, Cham (2020). https://doi.org/10.1007/978-3-030-65742-0_15
9. Krzanowski, W.: Between-groups comparison of principal components. J. Am. Stat. Assoc. **74**(367), 703–707 (1979)
10. Le Nguyen, T., Gsponer, S., Ilie, I., O'Reilly, M., Ifrim, G.: Interpretable time series classification using linear models and multi-resolution multi-domain symbolic representations. Data Min. Knowl. Disc. **33**(4), 1183–1222 (2019). https://doi.org/10.1007/s10618-019-00633-3
11. Lin, J., Keogh, E., Wei, L., Lonardi, S.: Experiencing SAX: a novel symbolic representation of time series. Data Min. Knowl. Disc. **15**(2), 107–144 (2007). https://doi.org/10.1007/s10618-007-0064-z
12. Löning, M., Bagnall, A., Ganesh, S., Kazakov, V., Lines, J., Király, F.J.: sktime: a unified interface for machine learning with time series. arXiv preprint arXiv:1909.07872 (2019)
13. Ruiz, A.P., Flynn, M., Large, J., Middlehurst, M., Bagnall, A.: The great multivariate time series classification bake off: a review and experimental evaluation of recent algorithmic advances. Data Min. Knowl. Disc. **35**(2), 401–449 (2020). https://doi.org/10.1007/s10618-020-00727-3
14. Satopaa, V., Albrecht, J., Irwin, D., Raghavan, B.: Finding a "kneedle" in a haystack: detecting knee points in system behavior. In: 2011 31st International Conference on Distributed Computing Systems Workshops, pp. 166–171. IEEE (2011)

15. Schäfer, P., Högqvist, M.: SFA: a symbolic Fourier approximation and index for similarity search in high dimensional datasets. In: Proceedings of the 15th International Conference on Extending Database Technology, pp. 516–527 (2012)
16. Schäfer, P., Leser, U.: Multivariate time series classification with WEASEL+ muse. In: ECML/PKDD Workshop on Advanced Analytics and Learning on Temporal Data (AALTD 2018), arXiv preprint arXiv:1711.11343 (2017)
17. Shokoohi-Yekta, M., Wang, J., Keogh, E.J.: On the non-trivial generalization of dynamic time warping to the multi-dimensional case. In: SDM (2015)
18. Singh, A., et al.: Interpretable classification of human exercise videos through pose estimation and multivariate time series analysis. In: 5th International Workshop on Health Intelligence (W3PHIAI 2021) at AAAI21. Springer (2021)
19. Yoon, H., Yang, K., Shahabi, C.: Feature subset selection and feature ranking for multivariate time series. IEEE Trans. Knowl. Data Eng. 17(9), 1186–1198 (2005)

Temporal Phenotyping for Characterisation of Hospital Care Pathways of COVID19 Patients

Mathieu Chambard[1], Thomas Guyet[2(✉)] [iD], Yên-Lan NGuyen[3], and Etienne Audureau[4] [iD]

[1] ENS Rennes/IRISA, Rennes, France
[2] Inria – Centre Grenoble Rhône Alpes, Lyon, France
thomas.guyet@inria.fr
[3] AP-HP, Hôpital Cochin, Sorbonne Université, INSERM UMR S 1138, Pierre Louis Institute of Epidemiology and Public Health, Paris, France
[4] AP-HP, Henri Mondor Hospital, University Paris Est Créteil, Créteil, France

Abstract. During the COVID19 crisis, Intensive Care Units admitted many patients with breathing disorders up to respiratory insufficiency. The care strategy of such patients was difficult to find and preventing patients to drift away toward a critical situation was one of the first challenge of physicians. In this study, we would like to characterize care pathways of patients that required a mechanical ventilation. The mechanical ventilation is an invasive treatment for the most critical respiratory insufficiencies. Through the analysis of the sequence of cares, we aim at supporting physicians to better understand patients evolution and let them propose new medical procedures to prevent some patients to be ventilated. This article proposes a method which combines a tensor factorization and sequence clustering. The tensor factorization enables to represent the care sequences as a sequence of daily phenotypes. Then, the sequences of phenotypes is clustered to extract typical care trajectories. This method is experimented on real data from Greater Paris university Hospital and is compared to a direct clustering of the sequences. The results show that the outputs are more easily interpretable with the proposed method.

Keywords: Tensor factorization · Sequence clustering · Phenotypes · Care pathways

1 Introduction

The advent of the COVID19 crisis show us the need to support physicians to identify as early as possible people who may have medical complications. This

This project is partly founded by Fondation APHP through the Chair AI-RACLES and received the agreement from the AP-HP CDW Scientific and Ethics Committee (CSE-20-11-COVIPREDS). Data used in preparation of this article were obtained from the AP-HP Covid CDW Initiative database. A complete listing of the members can be found at: (https://eds.aphp.fr/covid-19).

illustrates the need for predictive analytic tools that may support stakeholders to better manage crises in the future: better individual patient management, better patient flows organization, etc.

In our case study, we would like to characterize the pathways of patients that required mechanical ventilation. Mechanical ventilation is an invasive treatment indicated when the patient's spontaneaous breathing is inadequate to maintain effective gaz exchange. It is a heavy treatment that physicians try to avoid for their patients. Their characterization may help physicians to improve their medical management procedures in these cases [5].

The characterization of a patient suffering from a disease is often called a phenotype. A phenotype may be a collection of conditions (smoker status, comorbidities, BMI, treatments), but the notion of phenotype may be extended to recent procedures and drugs that have been delivered to the patient. The information of such procedure becomes a proxy for the patient condition.

In this work, we use Electronic Health Record (EHR) data from AP-HP (Greater Paris university hospital) to build phenotypes of patients. The data collected by information systems provide access to rich information on hospital stays and for a very large population of hospitalized patients. Then, the care trajectory of a patient is described as a matrix X with features (procedures or drugs) in columns and days in rows. The value $X_{i,j}$ is 1 when patient p received the procedure/drug i the day j. AP-HP has identified more than 20,000 patients hospitalized due to COVID19 from the beginning of the French crisis in March 2020 until March 2021.

There are potential flaws in the data but their volumetry and their sanitization make them valuable for extracting meaningful phenotypes. During the COVID19 crisis, physicians lack of time to code accurately the procedures or being exhaustive in their report. A sanitization of the database has been conducted all along the crisis to spur their use for research and operational purposes. These massive data should help to identify typical patient pathways, so called temporal phenotypes. A phenotype is a list of clinical features occurring in the same day for a subgroup of patients. For instance, the phenotype of patients suffering from a disease may be a combination of diagnosis codes, drugs or procedures he/she received, etc. A temporal phenotype describes a patient profile by the evolution of its "features" during his/her hospitalization. It groups together medications and procedures to best describe some visits.

This article proposes to use tensor factorization to identify automatically temporal phenotypes, so called typical care trajectories, from EHR data. More specifically, we investigate a simplified version of the CNTF model [14] which proposes to apply machine learning techniques in order to efficiently address tensor factorization (see Sect. 3). Our hypothesis is that depending on patients and procedures, their health status evolves in different ways. Discovering a temporal phenotype means to discover both what and when procedures occur during a patient stay and, if possible, to correlate the temporal phenotypes to patient outcomes such as mechanical ventilation. The dataset is presented in Sect. 4. Finally,

Sect. 5 presents and analyses the first results of our approach on COVID19 care pathways and is compared to KMeans clustering.

2   Related Works

Our goal is to discover phenotypes from an EHR database. Discovering phenotypes is an unsupervised task that aims at both describing phenotypes as typical sets of diseases and cares; and at identifying typical groups of patients having different types of phenotype.

There are several types of approach to address this problem. UPhenome [12] is a probabilistic approach based on Latent Dirichlet Allocation (LDA). It describes a patient by a set of cares without considering the temporal dimension. In our case study, we are interested in describing the longitudinal care trajectory of patients to characterize the dynamic of their disease. This dynamic of cares is characterized by careflow mining [3] using pattern mining techniques. In this approach, a careflow is a sequence of cares. But in case of sparse events, temporal patterns mining are more meaningful than sequential pattern mining. For instance, Dauxais et al. [4] proposed to discover patterns describing both the structural sequence of cares and the delay between. This problem has also been addressed in the statistical machine learning community. Many works have been proposed to discover structures in EHR data in supervised fashion. For instance, MedGraph [6] proposes a supervised EMR embedding method that captures the visit-code associations, and the temporal sequencing of visits through a point process.

In this article, we propose to explore an unsupervised statistical machine learning technique called non-negative tensor factorization (NNTF). NNTF has been studied extensively and many models have been proposed to tackle it [8]. The seminal methods are PARAFAC and Canonical Polyadic (CP) decompositions [7] which are the decomposition of a tensor in a finite collection of unidimensional vectors of rank R. The main limitation with this method is that it considers a tensor with fixed dimensions. In practice, it enforces all patients to have the same length of stay. Therefore, PARAFAC2 [9] extends the CP decomposition for a collection of matrices with different sizes (along one dimension). Both CP and PARAFAC2 are statistical approaches with good formal properties (e.g., uniqueness of the CP decomposition). Nonetheless, these approaches are not computationally tractable on large datasets. Recently, SPARTan [10] proposed an algorithmic reformulation of PARAFAC2 to be faster and more memory-efficient. Another way to solve the tensor factorization consists in using machine learning techniques that provide efficient approximated solving processes. Since the last years, several machine learning solutions for tensor factorization have been proposed with additional features, for instance temporal regularization [14], handling missing values [13] or optimized for sparse data [1, 2].

CNTF [14] (Collective Non-Negative Tensor Factorization) made two contributions: on the one hand, it is a flexible model which includes initial condition modeling, temporal regularization and classification regularization. Thus, CNTF is suitable for a wide range of care trajectories analysis. On the other hand, it proposes a description of a phenotype by a 2 dimensional matrices: one dimension for drugs and procedures and one dimension for lab tests. This matrix representation aims at capturing correlations between the two dimensions. Nonetheless, CNTF only enables to extract daily phenotypes, but not groups of entire care trajectories.

3   Care Pathway Characterization Through Tensor Factorization

In this section, we present a method for characterizing care pathways based on tensor factorization. The proposed method has two steps:

1. A tensor factorization identifies the daily phenotypes from patient care pathways,
2. The sequences of phenotypes are clustered to create groups of similar care pathways. The representative of each group is a *typical care trajectories*.

3.1   Tensor Factorization

In this section, we propose a factorization model inspired by CNTF [14]. Our model borrowed from CNTF the principle of tensor factorization through function minimization and the temporal regularization. We simplified the model by discarding the other constraints (including correlation modeling between lab tests and cares).

Tensor factorization is a data analysis technique that consists in decomposing a multidimensional tensor $\mathcal{X}$ into a collection of lower dimensional tensors $\mathcal{Y}_1, \ldots, \mathcal{Y}_k$ such that $\mathcal{X} \approx \mathcal{Y}_1 \otimes \cdots \otimes \mathcal{Y}_k$ where $\otimes$ is a matrix product. A nonnegative tensor factorization enforces $\mathcal{Y}$. matrices to contain only positive values.

In the context of EHR data analysis, $\mathcal{X}$ is seen as a three-dimensional tensor whose dimensions are the patient id (p patients), the time (d time units) and the medical events (N types of event). The length of stay of each patient visit are not all the same. Then, PARAFAC2 proposes a sparse representation of $\mathcal{X}$ as a collection of p two-dimensional matrices. I_k denotes the length of stay of the k-th patient such that its matrix $\boldsymbol{X}_k$ is of size $I_k \times N$.

Given $R \in \mathbb{N}$ the number of phenotypes, the matrix factorization problem consists in finding the matrices $\boldsymbol{U}$ of size $R \times N$ and the collections of p matrices $\boldsymbol{W}_k$ of size $I_k \times R$ such that for all $k \in \mathbb{N}_p$:

$$\boldsymbol{X}_k \approx \boldsymbol{W}_k \otimes \boldsymbol{U}$$

where $\boldsymbol{U}$ is the non-negative matrix describing the R phenotypes, and $\boldsymbol{W}_k$ is a non-negative matrix that describes the patient stay by the occurrence of the phenotypes each day. w_{krt} describes how likely the r-th phenotype exists at the particular time point t of patient k.

Inspired by CNTF, the problem is to minimize the following function:

$$f_{U,W_{1..p}} = \sum_{k=1}^{p} \frac{1}{I_k} \sum_{i,j} \hat{x}_{kij} - x_{kij} \log(\hat{x}_{kij})$$

where $\hat{X}_k = W_k \otimes U$ for all $k = 1..p$ is the tensor reconstruction from the phenotypes. In this problem formulation, the matrix reconstruction error is divided by the number of days. It aims at balancing contribution of patients who stayed for a long time or not.

At the moment, the temporal relationship is not taken into account in the model. However, for the course of a disease, we cannot look at the days independently of each other. The technique proposed in CNTF is to penalize a reconstruction model that does not allow to accurately predict the next sequences events or the stay outcomes. In both cases, we use a LTSM to model sequential dependencies between $w_{k \cdot t}$ vectors. The LTSM predicts the next state of the patient or the patient outcomes. In the first case, we want to minimize the mean square error between the real and predicted values, *i.e.*:

$$R(W_k) = \frac{1}{I_k} \sum_{t=2}^{I_k} ||g_k(w_{k \cdot (t-1)}) - w_{k \cdot t}||^2$$

where g denotes the prediction function of the LSTM trained on the sequence.

In the second case, we want to minimize the prediction error. In our practical case, the outcome of the stay is whether a patient has been mechanically ventilated or not. In such case, the error may be evaluated through the cross-entropy between the predicted and real outcomes.

Finally, the tensor factorization from EHR data is formalized by the following optimization problem:

$$\underset{U,W_{1..p}}{\arg\min} \ \mathcal{L} = \sum_{k=1}^{p} \frac{1}{I_k} \left(\sum_{i,j} \hat{x}_{kij} - x_{kij} \log(\hat{x}_{kij}) + \alpha \times \sum_{t=2}^{I_k} ||g_k(W_{t-1}) - W_t||^2 \right)$$

subject to $\hat{X}_k = W_k \otimes U$

$\qquad U \geq 0$

$\qquad W_k \geq 0, \ \forall k = 1..p$

where $\alpha \in \mathbb{R}^+$ is a parameter to balance the contribution of the two terms of the function.

To minimize $\mathcal{L}$, whatever optimization technique may be used. We use an alternating minimization strategy, illustrated in Algorithm 1. For each mini-batch B, the U is optimized given $W_{1..p}$ values, then $W_{1..p}$ is optimized using the U values. As the U is optimized several times per epoch while $W_{1..p}$ is optimized only once (for each batch, only one part of the matrix actually changes), then

---

Algorithm 1: Alternating minimization strategy (n epochs)

   Data: $X_{1..p}$: patient stays, R: the number of phenotypes

   Result: U: phenotypes, $W_{1..p}$, phenotype occurrences in patient stays

1 $U \leftarrow random$, $W_{1..p} \leftarrow random$;

2 **for** $e = 1..n$ **do**

3   **for** *Patient batch B* **do**

4        $U^* \leftarrow U + \frac{\partial f_{U,W_{k \in B}}}{\partial U}$;

5        $W^*_{k \in B} \leftarrow W_{k \in B} + \frac{\partial f_{U^*,W_{k \in B}} + \sum_{k \in B} R(W_k)}{\partial W_{k \in B}}$;

6     $U \leftarrow U^*$, $W_{1..p} \leftarrow W^*_{1..p}$;

---

we used different learning rates for each optimizer. In addition, the learning rate is decreased along the epochs to prevent from algorithm instability.

It is worth noting that we actually extract the phenotypes for the p patients. This means that the loss function $\mathcal{L}$ is evaluated on the p patients and splitting the datasets in train/test is not required.

3.2 Typical Care Trajectories

The tensor factorization enables us to change the representation of patient care pathways from sequences of cares $X_{1..p}$ to sequences of phenotypes $W_{1..p}$. In these two cases, the clustering of patients' matrices built typical care trajectories. It gathers similar pathways in clusters, and the representative of each cluster is a typical care trajectory.

In the general case, the patients' matrices do not have the same size due to different length of stay. Then, the classical clustering algorithms may not be applied. Our proposal is to use the Dynamic Barycentre Averaging (BDA) clustering approach [11]. DBA is a clustering algorithm for time series. It adapts the KMeans algorithm to the DTW distance. Thanks to the use of the DTW, it can cluster time series with different lengths. In our typical case, the sequence of phenotypes occurrences of a patient k, *i.e.* W_k, is seen as a multidimensional time series of length I_k and R dimensions. The centroid of a cluster computed by DBA is then a typical care trajectory.

In our experiments, all patients' stays have the same length. In this case, a simple KMeans algorithm applies for clustering the $W_{1..p}$ matrices.

4 Dataset of Ventilated COVID19 Patients

The objective of this study is to characterize the stays that have been admitted for COVID19. This disease affects the respiratory functions and may lead patients to critical respiratory insufficiency. In this case, patients have to be mechanically ventilated. This critical care procedure saves lives, but may lead to longer stays and to medical complications. For these reasons, physicians do their best to prevent patients from being mechanically ventilated.

In this section, we present the dataset that has been constructed to address the problem of the characterization of care pathways of patient who were ventilated.

Data were obtained from the AP-HP clinical data warehouse. It contains information for 27,370 ICU admissions with at least one positive PCR[1] test in one of the hospitals in the Greater Paris region between March 2020 and March 2021. It represents 17,400 unique patients. The database includes dates of admission to the intensive care unit, gender, age of each patient and possibly date of death.

For this study, patients were selected from people in the AP-HP database over 18 years old at ICU admission with a positive PCR test. We discarded patients having short visits (less than a day). In the original database there are 3.5 times more visits (27,370 visits) without ventilation procedure than visits leading to at least a mechanical ventilation procedure (6,066 visits). In order to balance the dataset, we subsample the patients without ventilation procedures. Indeed, the goal is to compare ventilated and non-ventilated patients. So the cohort must have roughly the same number of people and a similar age distribution. We adopted a stratified subsampling of the ventilated patients to have similar populations in age. More precisely, patients were drawn randomly to have for each age group (18–20, 20–40, 40–60, 60–80, 80–100, 100–120) as many ventilated as non-ventilated people. Figure 1, on the left, displays the age distributions of ventilated and non-ventilated patients. This figure also details the distributions of lengths of stay and of ages of death (for COVID19 or another reason). Note that in this study, we are not interested in the patient death but only on whether their stay leads to a mechanical ventilation or not.

The database contains medical and administrative information about each visit: clinical observations, lab test results, care performed or also textual medical reports. We decided to focus on medical procedures and prescription drugs, and to discard lab tests and medical reports. This information is collected with a suitable quality due to their administrative purpose (patient reimbursement). On the contrary, laboratory tests are too sparsely available and it is difficult to extract reliable information from medical reports.

All drugs and procedures delivered are timestamped and coded using standard taxonomies. Drugs are coded with ATC[2] codes and procedures are coded with CCAM[3] codes. CCAM is a French codification for medical procedures. Each code is a type of medical event in the $\mathcal{X}$ tensor. Drugs and procedure deliveries are timestamped with dates and times. We keep only the dates. For some procedures performed along several days (e.g., mechanical ventilation), the procedure is accurately recorded daily. Contrary to procedures, drugs are tagged with start and end dates, but the ends of drug exposures is not reliable. This is currently a potential weakness in our data.

---

[1] PCR (Polymerase chain reaction) denotes here a test for COVID19 infection.

[2] ATC: Anatomical, Therapeutical and Chemical.

[3] CCAM: Classification commune des actes médicaux/Common classification of medical procedures.

Table 1. Statistical characteristics of the cohorts/datasets. Raw database denotes the database of 21,901 patients with positive PCR tests, and final database denotes the stratified patients, with medical feature selection.

	Raw database	Final database
Number of patients	21,901	7,358
Number of visits	37,312	8,937
Average age	69 years	64 years
Gender distribution	M:56%, F:44%	M:62%, F:38%
Average length of stay	10 days	10 days
Number of different drugs	1,120	166
Number of different procedures	2,635	44
Death rate	23%	28%

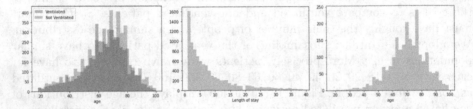

Fig. 1. Population characteristics. From left to right: age distribution, length of stay, age distribution of deceased people

The next step was to select a subset of potentially meaningful drugs/procedure among all possible codes. Indeed, the temporal and spatial complexity is exponential with the number of features. Considering the limited computational resources available on hospital servers, a selection of features was required. In addition, less medical features eases the interpretation of the results. The outputted phenotypes are more concise and there are less potential correlations to analyze for physicians.

In the case of the COVID19 study, the patients are very heterogeneous and have very different pathologies. The total number of medical events is very large, 1,120 different drugs and 2,635 different procedures. The selection of the medical features have been done in two steps. Firstly, the 500 most frequent drugs and 200 most frequent procedures were selected. Secondly, physicians selected 166 types of drugs and 44 types of procedure from the frequency-based selection. They selected the potentially most interesting medical features in the context of COVID19.

Table 1 sums up some characteristics of the cohort. Figure 1 on the right shows the distribution of the age of death. The distribution matches the known indicators: the people most affected are people over 60 years old.

Finally, for each visit, we select the events that occurs d days after the entry in an ICU. In case the patient visit started in another service, it is not taken into account. In this study, the pathway starts the first day in an ICU service. The entry date is used as an index date that is valid for patients who were ventilated or not. In addition, in the perspective of having a decision support tool, it is interesting to observe the care trajectory of a patient since its entry to decide as soon as possible the action to take to prevent a ventilation.

5   Experiments and Results on COVID19 Care Pathways

In this section, our method is applied to the database presented in the previous section. We remind that our objective is to investigate the care pathways of patients who have been mechanically ventilated or not. We set $d = 6$ meaning that 6 days were kept per patient from their arrival in ICU.

The tensor factorization model has been adapted from the CNTF implementation. It is implemented within the PyTorch framework. An initial study of the algorithm convergence shown that the algorithm does not significantly improve the results after 100 epochs. Then, we set the number of epochs to 100 and batch size of 100 patients. The running time on the dataset detailed in the next section is from 5 to 15 min on a server dedicated to AP-HP data analysis. This reasonable time makes the approach practical on real data. For the clustering algorithm, we use a K-means algorithm that suits our particular dataset which contains sequences of the same length. We used the K-Means *sklearn* library with a smart initialization of the centers.

In the remaining of this section, we start by studying the daily phenotypes extracted from care pathways of the whole dataset (ventilated and non-ventilated). Then, we investigate the results of the clustering phase of our method (typical care trajectories). Finally, we propose to compare the obtained results with the direct clustering of care trajectories.

5.1   Phenotypes of COVID19 Patients

The main parameters of our method are R, the number of phenotypes, and ρ, initial random state. Due to the stochastic nature of the optimization process, the results also depends on the initial random state (ρ). The method was tested with different $R \in [6, 12]$ and ρ in order to find which value to give to R and to ρ to have insightful and robust results.

In the following, we illustrate two cases: $R = 8$ and $R = 10$. The outputted phenotypes are illustrated in Fig. 2.

The detailed phenotypes are presented in Tables 2 and 3. After a physician's expertise, several pieces of information emerged from these phenotypes. First of all, we recognize phenotypes that characterize the pathway of patients in a intensive care unit. These are phenotypes with a prescription of thromboprophylaxis like *Enoxaparin* and also those who received antibiotics (*cefotaxime, amoxicillin*

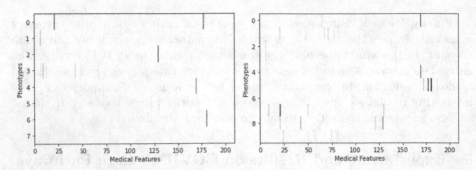

Fig. 2. Drugs phenotype result for $R = 8$ (on the left) and $R = 10$ (on the right). Each row corresponds to a phenotype, the columns correspond to drugs identifiers. A dark square means that the drug in column is part of the phenotype in row. The darker the square, the more likely the drug in this phenotype. (see Tables 2 and 3 for detailed values).

and inhibitor). This corresponds to a large part of the results: phenotypes 1.0, 1.2, 1.6, 1.7 and phenotypes 2.1, 2.2, 2.7, 2.8 and 2.9.

We also find deliveries of analgesics such as morphine, tramadol, nefopam or paracetamol in phenotypes 1.0, 1.2, 1.5, 1.7 and also in phenotype 2.7 and 2.8. These drugs treat muscle pain or fever caused by COVID19.

After some deaths from pulmonary embolism, a link has been discovered between a severe form of COVID19 and a risk of venous thrombosis. Patients gradually benefited from a preventive treatment for thrombosis such as *enoxaparin*. It appears in phenotypes 1.0, 1.2, 1.6, 2.1, 2.7 and 2.8.

The cohort has a high average age. This explains the appearance of *furosemide* in phenotype 1.3 and phenotype 2.8. This drug is an anti-hypertensive agent prescribed for elderly.

In addition, we observe common diseases in patients suffering from COVID19. First, some patients suffer from diabetes. Some have been intubated (phenotype 2.2), others have hypertension (phenotype 2.7: *amlodipine*). Second, some patients have cholesterol and cardiovascular problems. They are found in phenotypes 1.3 and 2.1. Finally, phenotypes 1.5 and 2.9 correspond to patients suffering from hypertension (*amlodipine, ramipril*) with also cardiovascular problems (*acetylsalicylic*).

Interestingly, the procedures are gathered in two or three phenotypes (1.4, 1.6, 2.0 and 2.4). Such phenotypes describes the standard monitoring procedures in a ICU service (*e.g.* electrocardiogram, intra arterial pressure). Thus, the stay of a patient being monitored in a ICU service is described with a combination of one of such phenotypes and phenotypes for drugs deliveries. It also highlight intubation procedures and the injection of *dobutamine/dopamine* present in phenotype 2.5.

Finally, it is worth noting that ρ parameter has a low impact on the results of the system. By repeating the experiment several times with different values, we observe similarities between the results of phenotypes. This robustness makes

Table 2. Phenotypes with R = 8. Numbers indicate the likelihood of the occurrence of a drug for a phenotype. Drugs names correspond to the French official denomination.

$Ph_{1.0}$	Enoxaparine: 1.2, Injection dobutamine/dopamine: 1.11, Paracetamol: 0.03, Dexamethasone: 0.0, Amlodipine: 0.0
$Ph_{1.1}$	Insuline aspart: 0.73, Monitoring of intra-arterial pressure: 0.03, Continuous monitoring of electrocardiogram: 0.0
$Ph_{1.2}$	Paracetamol: 2.33, Monitoring of intra-arterial pressure: 0.03, Enoxaparine: 0.03, Tramadol: 0.0, Acetylsalicylique acide: 0.0
$Ph_{1.3}$	Insuline glargine: 0.44, Insuline aspart: 0.24, Furosemide: 0.24, Atorvastatine: 0.18, Bisoprolol: 0.15
$Ph_{1.4}$	Continuous monitoring of electrocardiogram: 1.44, Central intra-arterial or intravenous pressure monitoring : 0.03, Monitoring of intra-arterial pressure: 0.03
$Ph_{1.5}$	Nefopam: 0.1, Acetylsalicylique acide: 0.09, Morphine: 0.09, Amlodipine: 0.08, Monitoring of intra-arterial pressure: 0.03
$Ph_{1.6}$	Central intra-arterial or intravenous pressure monitoring : 1.05, Monitoring of intra-arterial pressure: 0.06, Enoxaparine: 0.03, Acetylsalicylique acide: 0.0, Amlodipine: 0.0
$Ph_{1.7}$	Heparine: 0.12, Zopiclone: 0.12, Amoxicilline et inhibiteur d'enzyme: 0.09, Tramadol: 0.09, Nefopam: 0.06

us confident in the significance of the results. However, these are not exactly the same phenotypes. Sometimes a phenotype of an experiment is the mixture of two phenotypes of an experiment with a different value of ρ. This may be disturbing for physicians.

5.2   Care Trajectories

In this section, we describe the different pathways that lead to use mechanical ventilation or not. Then, we investigate the typical patient trajectories.

In the previous section, we analyzed the phenotypes, U. This section analyzes the information contained in $W_{1..p}$ matrices. These matrices represent the sequence of cares during the first 6 days of the ICU stay.

A cluster is a group of patients having the same kind of sequences during the first days of its stay. In our particular case, the clustering can be done with the DBA algorithm (see Sect. 3.2) or with a regular KMeans using the Froebenius distance between matrices having the same dimensions. For computational reasons, we applied this second alternative and set up the algorithm with $k = 6$.

Figure 3 illustrates the six cluster centers. For a better clarity, values lower than the half of the maximum of a matrix have been set to 0. A dark cell means that the phenotype is significantly present in average at a given day before starting ventilation for the group of patients.

The clusters could be split into three types of clusters. The clusters CT_0, CT_1 and CT_2 are mostly present in unventilated people. They are 2 to 3 times more

Table 3. Phenotypes with R = 10. (see legend of Table 2)

$Ph_{2.0}$	Continuous monitoring of electrocardiogram: 2.17, Monitoring of intra-arterial pressure: 0.03, Intubation trachéale: 0.03
$Ph_{2.1}$	Enoxaparine: 0.27, Dexamethasone: 0.22, Atorvastatine: 0.21, Bisoprolol: 0.18, Amoxicilline et inhibiteur d'enzyme: 0.15
$Ph_{2.2}$	Insuline aspart: 0.24, Zopiclone: 0.03, Monitoring of intra-arterial pressure: 0.03, Intubation trachéale: 0.03, Amoxicilline et inhibiteur d'enzyme: 0.03
$Ph_{2.3}$	Insuline aspart: 0.3, Phloroglucinol: 0.18, Insuline glargine: 0.15, Zopiclone: 0.15, Metformine: 0.06
$Ph_{2.4}$	Continuous monitoring of electrocardiogram: 1.96, Monitoring of intra-arterial pressure: 0.03, Intubation trachéale: 0.03
$Ph_{2.5}$	Central intra-arterial or intravenous pressure monitoring : 1.87, Injection dobutamine/dopamine: 1.29, Monitoring of intra-arterial pressure: 0.33, Intubation trachéale: 0.15, Continuous monitoring of electrocardiogram: 0.03
$Ph_{2.6}$	Continuous monitoring of electrocardiogram: 0.04, Prednisone: 0.03
$Ph_{2.7}$	Enoxaparine: 1.36, Insuline glargine: 0.38, Nefopam: 0.35, Amlodipine: 0.34, Ceftriaxone: 0.03
$Ph_{2.8}$	Paracetamol: 1.6, Furosemide: 0.55, Enoxaparine: 0.27, Morphine: 0.26, Tramadol: 0.12
$Ph_{2.9}$	Acetylsalicylique acide: 0.24, Cefotaxime: 0.15, Prednisone: 0.15, Amlodipine: 0.12, Ramipril: 0.09

Table 4. Repartitions of ventilated/unventilated patients.

Care trajectories	Patients	Unventilated	Ventilated
CT_0	362	257	105
CT_1	3093	2110	983
CT_2	2889	1376	1513
CT_3	883	640	243
CT_4	767	30	737
CT_5	884	1	883

present in non-ventilated patients than in ventilated patients. Then the clusters CT_4 and CT_5 are especially present in ventilated people. Finally, cluster CT_3 lies in both visits from ventilated and non-ventilated people.

We remind that the hospital stay of patients is aligned with the first days of hospitalization. Therefore, we can have a shift in phenotypes between patients depending on their health status at arrival. This shift is observed with cluster CT_1 and CT_3. These two clusters have almost the same phenotypes: 2.1, 2.2,

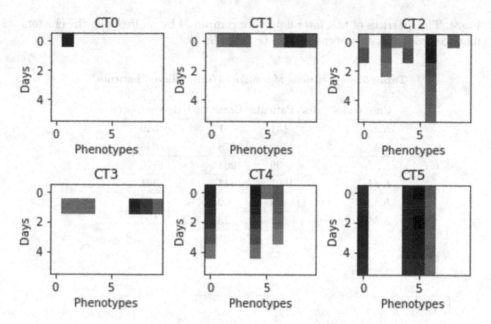

Fig. 3. Typical care trajectories (CT_i for $i = 1..6$) of patient during the first 6 days of hospitalization. A typical care trajectory gives how likely a phenotype appears at a given day of the stay. This figure uses the phenotypes extracted with $R = 10$ (see Fig. 2, on the right).

2.3, 2.7, 2.8, 2.9. One is filled on the first day of hospitalization, the other on the second day. In addition, the clusters have the same proportion of ventilated and unventilated, which supports the fact that these clusters represent the same kinds of patients.

The phenotypes 2.0, 2.4, 2.5 and 2.6 are mostly present in ventilated patients. They correspond to cluster CT_4 and CT_5 for which 98% of the patients have been ventilated. Indeed, the phenotypes contained in the clusters are phenotypes linked to classical resuscitation procedures and are very similar to the ones in the previous section.

Finally, cluster CT_2 represents as many ventilated and non-ventilated patients. This cluster appears in almost a third of patients. It is made up of phenotypes 2.0, 2.4 and 2.6 which are mainly made of ICU procedures. The other phenotypes present are phenotypes 2.2 and 2.6 which are mainly prescriptions.

5.3   Comparison with Direct Clustering

In this section, the goal is to compare the trajectories extracted with our method and the ones extracted with a direct clustering (K-Means). Figure 5 shows the KMeans cluster centers. It illustrates the medical events' occurrences wrt days. The clusters of this figure are compared to the results of our method presented in

Fig. 4. The matrices of this later figure are computed by multiplying the clusters matrices with the phenotype matrix U (see Fig. 2).

Table 5. Repartitions of ventilated/unventilated patients.

Care trajectories	Patients	Unventilated	Ventilated
KM_0	541	1	540
KM_1	2422	1778	644
KM_2	3839	1963	1876
KM_3	1005	644	361
KM_4	450	0	450
KM_5	621	28	593

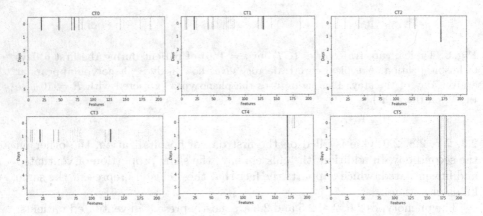

Fig. 4. Typical care trajectories: medical events along the first 6 days of hospitalization (alternative view of the result presented in Fig. 3)

Table 5 provides the number of ventilated and non-ventilated patients in each cluster.

We can observe common clusters between the two method results. For instance, there is a strong similarity between CT_1 and KM_3. Additionally, CT_4 and CT_5 clusters look like KM_0, KM_4 and KM_5 clusters. Then, we can conclude that the approaches extract almost the same care trajectories.

Nonetheless, KM_3 and KM_4 are quite similar while there is more diversity in the phenotypes extracted by our methods. A possible explanation is that clustering the sequence of few phenotypes is easier than clustering the sequence of all the medical events.

The second advantage of our method is in the ease to interpret the results and get insight from them. We have seen that the daily phenotypes can be

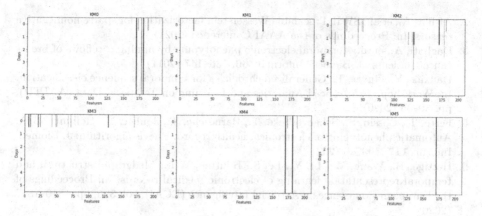

Fig. 5. Typical care trajectories (K-Means clustering): medical events along the first 6 days of hospitalization.

interpreted by physicians. This intermediary interpretation enables physicians also to get insights from the typical care trajectories of Fig. 2. We believe a direct clustering providing the care trajectories without intermediary phenotype is harder to interpret.

6 Conclusion

We presented a method to extract typical care trajectories from EHR care pathways. Our method combines a tensor factorization to extract daily phenotypes and a clustering of phenotype sequences. This method has been applied to the analysis of COVID19 patients admitted in ICU to investigate the use of mechanical ventilation.

The first results with this method are interesting. First, the use of an approximate tensor factorization inspired by CNTF enables to process a large number of patient sequences. Phenotypes have been easily interpreted by physicians as their evolution over days. Compare to the direct clustering of the sequences, we argue that the use of phenotype is more insightful and easier to interpret. Finally, these results are promising. It is important to continue to look at the evolution of phenotypes in patients to compare the course of the disease in different subgroups of the population. For the future, the goal will also be to compare the evolution in ventilated and non-ventilated people using supervised tensor factorization techniques.

References

1. Afshar, A., Perros, I., Papalexakis, E.E., Searles, E., Ho, J., Sun, J.: COPA: constrained PARAFAC2 for sparse & large datasets. In: Proceedings of the 27th ACM International Conference on Information and Knowledge Management, pp. 793–802 (2018)

2. Afshar, A., et al.: SWIFT: scalable Wasserstein factorization for sparse nonnegative tensors. In: Proceedings of the AAAI Conference (2021)
3. Dagliati, A., et al.: Temporal electronic phenotyping by mining careflows of breast cancer patients. J. Biomed. Informat. **66**, 136–147 (2017)
4. Dauxais, Y., Guyet, T.: Generalized chronicles for temporal sequence classification. In: Workshop on Advanced Analytics and Learning on Temporal Data (AALTD), pp. 30–45 (2020)
5. Ferté, T., Cossin, S., Schaeverbeke, T., Barnetche, T., Jouhet, V., Hejblum, B.P.: Automatic phenotyping of electronical health record: Phevis algorithm. J. Biomed. Inform. **117**, 103746 (2021)
6. Hettige, B., Wang, W., Li, Y., Le, S., Buntine, W.L.: Medgraph: structural and temporal representation learning of electronic medical records. In: Proceedings of the European Conference on Artificial Intelligence (ECAI), vol. 325, pp. 1810–1817 (2020)
7. Hitchcock, F.L.: The expression of a tensor or a polyadic as a sum of products. J. Math. Phys. **6**(1–4), 164–189 (1927)
8. Hong, D., Kolda, T.G., Duersch, J.A.: Generalized canonical polyadic tensor decomposition. SIAM Rev. **62**(1), 133–163 (2020)
9. Kiers, H.A., Ten Berge, J.M., Bro, R.: PARAFAC2-part I. A direct fitting algorithm for the PARAFAC2 model. J. Chemom. Soc. **13**(3–4), 275–294 (1999)
10. Perros, I., et al.: Discovery and data mining -ACM SIGKDD), pp. 375–384 (2017)
11. Petitjean, F., Ketterlin, A., Gançarski, P.: A global averaging method for dynamic time warping, with applications to clustering. Patt. Recogn. **44**(3), 678–693 (2011)
12. Pivovarov, R., Perotte, A.J., Grave, E., Angiolillo, J., Wiggins, C.H., Elhadad, N.: Learning probabilistic phenotypes from heterogeneous EHR data. J. Biomed. Inform. **58**, 156–165 (2015)
13. Yin, K., Afshar, A., Ho, J.C., Cheung, W.K., Zhang, C., Sun, J.: LogPar: logistic PARAFAC2 factorization for temporal binary data with missing values. In: Proceedings of the International Conference on Knowledge Discovery & Data Mining (ACM SIGKDD), pp. 1625–1635 (2020)
14. Yin, K., Qian, D., Cheung, W.K., Fung, B.C.M., Poon, J.: Learning phenotypes and dynamic patient representations via RNN regularized collective non-negative tensor factorization. In: Proceedings of the AAAI Conference on Artificial Intelligence, vol. 33, pp. 1246–1253 (2019)

Non-parametric Multivariate Time Series Co-clustering Model Applied to Driving-Assistance Systems Validation

Etienne Goffinet[1,2]([⊠]), Mustapha Lebbah[1], Hanane Azzag[1], Giraldi Loïc[2], and Anthony Coutant[1]

[1] Sorbonne Paris-Nord University, LIPN-UMR, 7030 99 Avenue Jean Baptiste Clément, Villetaneuse, France
etienne.goffinet@lipn.univ-paris13.fr
[2] Groupe Renault SAS, Avenue du Golf, Guyancourt, France

Abstract. In this paper, we propose a new Bayesian co-clustering approach applied to Multivariate time series. Our methodology of Functional Non-Parametric Latent Block Model (FunNPLBM) simultaneously creates a partition of observation and a partition of temporal variables, using latent multivariate gaussian block distributions. We propose to use a bi-dimensional Dirichlet Process as a prior for the block distributions parameters and for block proportions, which natively provides model selection. This approach is benchmarked and studied on a simulated dataset and applied to an advanced driver-assistance system validation use-case.

Keywords: Bayesian non parametric · Co-clustering · Model-based clustering · Multivariate time series · Driving-assistance systems

1 Introduction

Unsupervised classification, or *clustering*, is a first approach to dataset exploration that consists in the automatic grouping of similar observations into homogeneous groups, without supervision, i.e., without labels. Time series clustering is crucial to decision-making in many domains (Industry, Health, Finance, Biology,...) and has been extensively studied in the literature [1,2].

In a multivariate setting, the clustering methods deal with several variables simultaneously. The *Co-clustering* (also called Bi-clustering, or block clustering) simultaneously produces a partition of observations and a partition of variables (respectively *row-partition* and *column-partition*). This approach creates a structure that highlights the dependencies between observation groups and variables distributions. The co-clustering has been applied in various fields, (e.g., genetics [27], biological applications [38], text mining [45]) and has been addressed with a large numbers of methods: through spectral analysis [12], matrix factorization [31], information theory [13], and, more recently, optimal transport [30] and deep learning [44]. The Latent Block Model (LBM) [19] is a model-based co-clustering

© Springer Nature Switzerland AG 2021
V. Lemaire et al. (Eds.): AALTD 2021, LNAI 13114, pp. 71–87, 2021.
https://doi.org/10.1007/978-3-030-91445-5_5

method, recently used in several domains [21,25]. The model-based approach natively provides missing values inference and probabilistic outliers detection, while keeping a sparse parameter number, which helps interpretability.

In the standard LBM approach, datasets are composed of 1-d cells, i.e., the considered dataset is a matrix, and the co-cluster (or *block*) distributions are univariate. LBM methods for time series, where each cell is a temporally-indexed vector, have only been introduced recently. The method FLBM from [9] uses a piecewise polynomial regression model as block distribution, which assumes that every time series admits a latent segmented structure in a common polynomial basis.

However, FLBM does not reduce the time series dimension as it directly models the time series in the observation space and is impractical for high-dimensional time series datasets. The method FunLBM [6], by contrast, includes a dimension reduction step, which relies on functional PCA projections [36] of the time series, and assumes a multivariate Normal model for the block distribution. In a different context, some works extend these approaches by assuming the presence of several independent [18] or hierarchically nested [10] partitions.

A limitation of FunLBM is that, as a parametric model, the number of blocks is assumed known a priori. In practice, it is rarely true, and this number must be estimated with an additional model selection step. This selection is usually performed either with a grid-search or by hierarchically exploring existing clusters with greedy optimization. These strategies have drawbacks: 1) with the grid search, the computation cost can be prohibitive as every combination of block number is tested, and the user is never certain that the true model is within the grid; 2) the greedy optimization heuristic is sub-optimal, by picking iteratively local optima and assuming a hierarchical structure of the mixture components; 3) whether with the greedy optimization or grid-search, choosing the model selection criterion [8,15] is not an easy task and influences the final results.

The Dirichlet Process Mixture Model (DPMM) is a Bayesian non-parametric model-based clustering approach that can infer the number of latent clusters. As a *non-parametric* model, its parameter set dimension may increase indefinitely with the dataset size. This model is particularly well suited to massive dataset exploration, especially when it is possible to allocate additional resources to augment the dataset and discover new observation space areas.

Non-parametric approaches to LBM (NPLBM) have been studied in few works [22,32], but, to the best of our knowledge, never applied to multivariate time series co-clustering. The non-parametric framework has also been used by [5] for the clustering of multivariate time series data based on a grid-based multivariate density estimation, but not for coclustering. This paper proposes to close the gap between FunLBM and NPLBM with *functional non-parametric Latent Block Model* (funNPLBM), the first non-parametric model-based method applied to multivariate time-series co-clustering.

In addition, our contribution includes a practical use case illustrating the method's capacities, a more compact definition of the NPLBM, experiments, hindsight on the inference settings, and Scala code provided for reproducibility.

This paper illustrates the funNPLBM application to advanced driver-assistance systems (ADAS) development, which remains a challenge for car manufacturers. These systems (e.g., emergency braking, lane centering,...) are introduced gradually into new cars. Given the high number of car models, driving conditions, traffic laws, and given the expected reliability, it is today impossible to validate ADAS rigorously with only physical "on-tracks" tests. *Groupe Renault* has invested in massive driving simulation technology to circumvent this issue. The simulation tool mimics car driving conditions based on vehicle physics, driver behavior, and interaction with a configurable environment. The simulation outputs a large amount of information, mainly as multivariate time series with unequal length. Simulated datasets dimensions are considerable: for a given use case, the number of simulations can be as large as $\mathcal{O}(10^6)$, with $\mathcal{O}(10^3)$ variables, each recording $\mathcal{O}(10^4)$ time steps. Overall, more than $\mathcal{O}(10^{13})$ data points are produced.

In the following, Sect. 2 presents a review of the model-based clustering and co-clustering. Section 3 describes funNPLBM, its inference, and the time series preprocessing. Benchmark and experiments are studied in Sect. 4 and, finally, a real-case application on an industrial dataset is presented in Sect. 5.

2   Related Work

In the next sections we use the following notations: $X = (x_{i,j,s})_{n \times p \times d}$ is the dataset, where n is the number of rows (in our cases, the simulation number), p the number of columns (in our case, the number of simulated variables) and d the observation space dimension (c.f. time series dimension reduction step in introduction of Sect. 3.1).

We denote $\mathbf{x}_{i,.} = (x_{i,j})_{1 \leq j \leq p}$ the i-th row of X, $\mathbf{x}_{.,j} = (x_{i,j})_{1 \leq i \leq n}$ the j-th column. $X_{-i,.}$ and $X_{.,-j}$ designates the dataset without the corresponding row or column. The row-clusters memberships vector is noted $\mathbf{z} = (z_i)_n$ and the column-clusters memberships $\mathbf{w} = (w_j)_p$, such that $(z_i, w_j) = (k, l)$ indicates that element $x_{i,j}$ belongs to row-cluster k and column-cluster l.

2.1   Model-Based Clustering and Dirichlet Process Mixture Model

Mixture Model (MM) [11] is a probabilistic clustering approach which assumes that the overall density on the $(p.d)$-dimensional space is a convex combination of densities: $p(\mathbf{x}_{i,.}) = \sum_{k=1}^{K} \pi_k F(\theta_k)$, with $\pi_k = p(z_i = k)$, and $F(\theta_k) = p(\mathbf{x}_{i,.}|z_i = k)$ is the distribution of $\mathbf{x}_{i,.}$ in component k, with density family F. With this definition, sampling $\mathbf{x}_{i,.}$ is performed by first drawing a membership $z_i \sim Mult(\pi)$ then drawing from $F(\theta_{z_i})$. Model inference is performed by likelihood optimization using the Expectation-Maximization (EM) algorithm [11]. The MM admits the alternative representation:

$$\forall i \in \{1,..,n\}, \; \mathbf{x}_{i,.} \mid \theta_i \sim F(\theta_i), \; \theta_i \sim G, \; G = \sum_{k=1}^{K} \pi_k \delta_{\theta_k},$$

with δ_θ the Dirac delta function, In this definition, each observation x_i is associated to a parameter θ_i. Because G is finite and $K < n$, several θ_i are similar, which creates groups of elements with common distribution. The Dirichlet Process Mixture Model (DPMM) can be seen as a Bayesian non-parametric extension of the MM where G is now an infinite random distribution with a Dirichlet Process (DP) prior distribution. This prior is a distribution over distribution that takes two parameters: a concentration α and a base distribution G_0. The distribution G admits [40] the stick-breaking representation $G = \sum_{k=1}^{\infty} \pi_k(\mathbf{v}) \delta_{\theta_k}$, with

$$\pi_i(\mathbf{v}) = v_i \prod_{j=1}^{i-1} (1 - v_j), \quad \mathbf{v} = (v_i)_{\{1,\ldots,n\}}, \quad v_j \overset{\text{i.i.d.}}{\sim} \text{Beta}(1, \alpha), \quad \theta_j \overset{\text{i.i.d.}}{\sim} G_0.$$

Several methods have been developed to infer DPMM's parameters, either based on variational inference [4], or Markov chain Monte Carlo (MCMC) [14,34]. For large datasets applications, variational inference methods have often been preferred over MCMC for their speed, at the cost of hypothesis on the posterior distribution structure. However, recent works [33,42] have made MCMC processes scalable and rehabilitate their use for industrial purposes. In particular, the collapsed Gibbs sampler is a natively fast MCMC's method that assumes the prior distribution G_0 conjugate to the density family F. This assumption enables close-form computations of the prior and posterior predictive distributions, that are used to estimate posterior cluster membership probabilities.

2.2   Latent Block Model

The LBM [21] is a bi-dimensional MM, that assumes the presence of a finite number of latent column-clusters in addition to the observation partition. Inside a block $X_{k,l}$, each cell follows the component distribution $F(\theta_{k,l})$, with F a density family. The LBM likelihood of X is given by:

$$p(X) = \sum_{\mathcal{Z} \times \mathcal{W}} p(\mathbf{z}, \mathbf{w}) p(X \mid \mathbf{z}, \mathbf{w}) = \sum_{\mathcal{Z} \times \mathcal{W}} p(\mathbf{z}) p(\mathbf{w}) p(X \mid \mathbf{z}, \mathbf{w}),$$

where $\mathcal{Z}$ and $\mathcal{W}$ respectively denote the sets of all possible row and column partitions. The row-membership distribution $p(\mathbf{z})$ is defined as $\prod_i p(z_i) = \prod_i \pi_{z_i}$, with $\pi = (\pi_k)_K$ the mixing proportions, and $p(\mathbf{w}) = \prod_j p(w_j) = \prod_j \rho_{w_j}$. The density of X is $p(X \mid \mathbf{z}, \mathbf{w}) = \prod_{k,l} \prod_{x \in X_{k,l}} p(x \mid \theta_{k,l})$, with $x \sim F(\theta_{k,l})$. The inference process is usually performed in an Expectation-Maximization fashion, e.g. with a Stochastic-Gibbs [26] approach, or variational inference [20]. In the following, we define the funNPLBM and the stochastic inference process.

3   Functional Non-parametric Latent Block Model

This section introduces the novel *Functional Non-Parametric Latent Block Model* and the inference. Because multivariate time series are observed in high-dimensional spaces, in which models are known to suffer from the *curse of dimensionality*, it is essential to preprocess the dataset with a dimension reduction method. In addition, this step greatly reduces the computation cost. The functional PCA (fPCA) [36] is a two-step dimensionality reduction method, popular in the parametric model-based setting [6,41]. This method handles time series with unequal lengths, as is the case in our applications. During fPCA, the time series are first projected in a common polynomial basis. Among the many candidate representations available in the literature (e.g., Fourier, Legendre, Chebyshev, ...), we use an interpolated log-scaled Fourier periodogram, as advocated in [7]. This transformation consists in projecting each time series individually in the frequency domain, then interpolating the obtained log-periodograms in a common frequency basis. After this transformation, the second step of the fPCA consists in projecting the obtained coefficients in a lower-dimensional space using PCA. As a result, the unequal-length time series, described by $\mathcal{O}(10^3)$ points at the origin, are transformed into equal-length d-dimensional vectors, with d the number of PCA axes kept (usually less than 10).

3.1 Functional Bayesian Non-parametric Latent Block Model

In another context than the multivariate time series analysis, [32] proposed a definition of the NPLBM. This work assumes Pitman-Yor Process (PYP) priors for the row-memberships and column-membership. However, PYP priors (as DP priors) are distributions over parameters and not on memberships. Using PYP, the authors are in fact implicitly defining sets of parameter distributions that are not linked to the block distributions. In the following, we propose a comprehensive definition that links block distributions and memberships intuitively.

This definition is based on a bi-dimensional extension of the DP. Each dataset cell $x_{i,j}$ is associated with a parameter $\theta_{i,j}$, grouped in the $n \times p$ matrix Θ. Two Dirichlet Process priors are used: one on Θ's rows and one on Θ's columns. This double definition ensures that every cells of a row belongs to the same row-cluster and every column element to the same column-cluster. This process is defined by:

$$x_{i,j} \mid \theta_{i,j} \sim F(\theta_{i,j})$$

$$\theta_{i,.} \mid G \sim G, G = \sum_{k=1}^{\infty} \pi_k(\mathbf{s}) \, \delta_{\theta_{k,.}},$$

$$\theta_{.,j} \mid H \sim H, H = \sum_{l=1}^{\infty} \rho_l(\mathbf{t}) \, \delta_{\theta_{.,l}},$$

$$\pi_k(\mathbf{s}) = s_k \prod_{k'=1}^{k-1} (1 - s_{k'}), \quad \mathbf{s} = (s_k)_{\{1,\dots,n\}}, \quad s_k \overset{\text{i.i.d.}}{\sim} \text{Beta}(1, \alpha),$$

$$\rho_l(\mathbf{t}) = t_l \prod_{l'=1}^{l-1} (1 - t_{l'}), \quad \mathbf{t} = (t_l)_{\{1,\dots,p\}}, \quad t_l \overset{\text{i.i.d.}}{\sim} \text{Beta}(1, \beta).$$

With this definition, generating a dataset X is done by drawing π and ρ, sampling $\mathbf{z}$ and $\mathbf{w}$ separately, then sampling Θ given $\mathbf{z}$ and $\mathbf{w}$, and finally drawing the cells value $x_{i,j}$ from $F(\theta_{i,j})$. The likelihood of X is given by $p(X \mid \mathbf{z}, \mathbf{w}, \Theta) = \prod_{i,j} p(x_{i,j} \mid \theta_{i,j})$, and the joint prior distribution of the $\mathbf{z}$ and $\mathbf{w}$ is given by:

$$p(\mathbf{z}, \mathbf{w}, \mathbf{t}, \mathbf{s}, \Theta \mid G_0, \alpha, \beta) = p(\mathbf{z} \mid \mathbf{s})\, p(\mathbf{s} \mid \alpha)\, p(\mathbf{w} \mid \mathbf{t})\, p(\mathbf{t} \mid \beta)\, p(\Theta \mid G_0).$$

In the next section we describe the bi-dimensional stochastic inference process.

3.2   Model Inference

The inference is performed with a collapsed Gibbs sampler that simulates draws from the posterior distribution $p(\mathbf{z}, \mathbf{w} \mid X, G_0, \alpha)$. This approach directly uses the predictive distributions closed form and therefore does not require sampling block parameters. At each iteration m, the sampler alternates the following two-steps:

$$1.\ \text{Draw } \mathbf{z}^{(m+1)} \mid \mathbf{w}^{(m)}, X, \alpha, G_0,$$
$$2.\ \text{Draw } \mathbf{w}^{(m+1)} \mid \mathbf{z}^{(m+1)}, X, \beta, G_0.$$

During the first step, the row memberships update is performed sequentially: each row-cluster membership z_i is updated with the other row-memberships $\mathbf{z}_{-i}$ and column-partition $\mathbf{w}^{(m)}$ fixed, following $p(z_i = k \mid \mathbf{z}_{-i}, X, \mathbf{w}^{(m)}, \alpha, G_0) \propto$

$$\begin{cases} \dfrac{n_k}{n-1+\alpha}\, p(\mathbf{x}_{i,.} \mid \mathbf{w}^{(m)}, X_{-i}, \mathbf{z}_{-i}, G_0), & \text{existing cluster } k, \quad (1) \\[3mm] \dfrac{\alpha}{n-1+\alpha}\, p(\mathbf{x}_{i,.} \mid \mathbf{w}^{(m)}, G_0), & \text{new row-cluster}, \quad (2) \end{cases}$$

where n_k is row-cluster k size. We emphasize that the parameters Θ do not appear in these formulas, as they are integrated over in the predictive distributions.

In Eq. (2), $p(\mathbf{x}_{i,.} \mid \mathbf{w}^{(m)}, G_0) = \prod_l p(\mathbf{x}_{i,l}^{(m)} \mid G_0)$, with $\mathbf{x}_{i,l}^{(m)}$ the elements of row i in column-cluster l at iteration m. For each column-cluster l, the prior predictive distribution of $\mathbf{x}_{i,l}^{(m)}$ is obtained by integrating over the component's parameter: $p(\mathbf{x}_{i,l}^{(m)} \mid G_0) = \int_\theta p(\mathbf{x}_{i,l}^{(m)} \mid \theta)\, p(\theta \mid G_0)\, d\theta$. Because G_0 is a prior conjugate to F, this integral is analytically tractable (see Sect. 3.3 for the detail in the multivariate gaussian case). The joint posterior predictive distribution $p(\mathbf{x}_{i,.} \mid \mathbf{w}^{(m)}, X_{-i}, G_0)$ from Eq. (1) has the same definition, with G_0 updated with the observations inside the blocks.

The second step of the Gibbs-sampler is performed symmetrically on column clusters. Once the maximum number of iterations reached, the row and column final partitions are estimated with the mean of the partitions sampled after burn-in (c.f. Sect. 3.4 –Sect. 3 for computation detail). The algorithm global complexity is linear in n, p, in the current blocks number and in the iterations number, but also depends on the complexity of the sufficient statistics update and predictive distribution computation. The inference is summarized in Algorithm 1. In the next subsection we detail how Eqs. (1) and (2) simplify with our choice of G_0.

3.3   Multivariate Gaussian Case

After the time series preprocessing step, each dataset cell $x_{i,j}$ is a d-dimensional numeric vector produced by the fPCA, that we model with a multivariate Gaussian density. As as conjugate prior, we choose G_0 to be the Normal Inverse Wishart (NIW) distribution with hyper-parameters $(\mu_0, \lambda_0, \Psi_0, \nu_0)$. Given $X_{k,l}$, the observations in block (k,l), the block parameters posterior distribution is formally defined by $\mathrm{p}(\mu, \Sigma \mid X_{k,l}) = NIW(\mu, \Sigma \mid \mu_{k,l}, \kappa_{k,l}, \Psi_{k,l}, \nu_{k,l})$, with:

$$\mu_{k,l} = \frac{\kappa_0 \mu_0 + n_{k,l}\overline{x}_{k,l}}{\kappa_{k,l}}, \quad \kappa_{k,l} = \kappa_0 + n_{k,l}, \quad \nu_{k,l} = \nu_0 + n_{k,l},$$

$$\Psi_{k,l} = \Psi_0 + C + \frac{\kappa_0 n_{k,l}}{\kappa_{k,l}}(\mu_0 - \overline{x}_{k,l})(\mu_0 - \overline{x}_{k,l})^T, \quad C = \sum_{x \in X_{k,l}} (x - \overline{x}_{k,l})(x - \overline{x}_{k,l})^T.$$

With these parameters and following [16], the joint posterior predictive distribution needed in Eq. (1) is the multivariate t-student distribution:

$$t_{\nu_{k,l}-p+1} \left(\mu_{k,l}, \frac{(\kappa_{k,l}+1)\,\Psi_{k,l}}{\kappa_{k,l}(\nu_{k,l}-p+1)} \right).$$

This definition outlines the cubic complexity in d, due to the t-student density computation cost. The next section details the inference process implementation.

3.4   Implementation

G_0 **Hyperparameters Specification.** The clustered objects are PCA coefficients, which are centered. Therefore we set μ_0 to be the d-dimensional zero vector. The precision matrix Ψ_0 specification is a bit trickier and depends on assumptions on the dataset. For non-parametric autoregressive models, [39] compares several prior specifications for Ψ_0 and concludes that the dataset precision obtained with maximum likelihood estimation is a good standard, which we keep in our application. κ_0 and ν_0, which represent the user's confidence in μ_0 and Ψ_0, are set to their lowest value, as we want them as uninformative as possible.

Initialization Strategy. Initializing a bayesian non-parametric MCMC inference algorithm is often done with single-component partition [35,37]. However, [23] shows that this strategy may be suboptimal when dealing with high-dimensional datasets and high component numbers, and recommends to use random partition as initial state, with more components than the actual component number. However, this number is unknown. A tempting solution is to initialize the partitions with one component per observation, but this choice is computationally expensive because the membership update has linear complexity in the block number. We propose a heuristic, consisting of running the inference process twice. In a first run, the inference is initialized with a one-cluster partition; after this first run completion, the maximum block number sampled during the inference is kept and used as the initial number of components for the second run.

Infering the Final Partitions. In output of the Gibbs sampler algorithm, the user gets a set of sampled row-partitions $(\mathbf{z}^{(m)})_m$ and column-partitions $(\mathbf{w}^{(m)})_m$, that must be aggregated to obtain the final partition modes: $\hat{\mathbf{z}}$ and $\hat{\mathbf{w}}$. This *consensus partition* estimation is an NP-complete problem [29], that several works have addressed (c.f. [43] for an extensive review). We use the recent method [17], which proposes an efficient extension of a combinatorial optimization method [24] that construct the partition with the minimal distance [28] to the samples, without assumptions on the final number of clusters or on the clustering structure.

Algorithm 1: FunNPLBM Inference

---

input : Dataset X, $n \times p \times d$ tensor

$\quad\quad\quad$ α, β, G_0 (c.f. specification strategies in Sect. 3.4–Sect. 1)

$\quad\quad\quad$ Iteration number M

output: Estimated row-partition $\hat{\mathbf{z}}$ and column-partition $\hat{\mathbf{w}}$

Initialize $\mathbf{z}^{(0)}$ and $\mathbf{w}^{(0)}$ (c.f. initialization methods in Sect. 3.4–Sect. 2)

for $m \leftarrow 1$ **to** M **do**

$\quad$ **for** $i \leftarrow 1$ **to** n **do**

$\quad\quad$ Compute $p(z_i \mid \mathbf{z}_{-i}, X, \mathbf{w}^{(m)}, \alpha, G_0)$ as defined by Eqs. (1) and (2)

$\quad\quad$ Sample $z_i^{(m+1)}$

$\quad$ **for** $j \leftarrow 1$ **to** p **do**

$\quad\quad$ Compute $p(w_j \mid \mathbf{w}_{-j}, X, \mathbf{z}^{(m+1)}, \beta, G_0)$ as defined by Eqs. (1) and (2)

$\quad\quad$ Sample $w_j^{(m+1)}$

Average the partitions (c.f. Sect. 3.4 - §3) to obtain the final partitions $\hat{\mathbf{z}}$ and $\hat{\mathbf{w}}$.

---

4 Experiments on Synthetic Data

4.1 Experimental Setup

Benchmark and experiments are conducted on a dataset sampled from a known generative model. The dataset is generated by sampling from the distributions $\mathcal{N}\left(f_{k,l}(t), s^2\right)$ where $f_{k,l}$ is a given *prototype* function and $s = 0.02$. The estimated block partition quality is compared to the known generative partition, based on several scores: the Rand Index (RI), Adjusted Rand Index (ARI) and the Normalized Mutual Information (NMI). The RI is a popular criterion choice in the clustering domain, which represents the proportion of correctly grouped and separated observations with respect to the observed classes. The ARI is a corrected-for-chance version of the RI that takes into account the probability of getting good RI at random. The NMI is an entropy-based criterion from the information theory literature estimating the quantity of knowledge a partition gives on another. In the following benchmark and experiments, we work on a dataset of dimension 140×140, with unbalanced row cluster sizes $(20, 30, 40, 30, 20)$ and column cluster sizes $(40, 20, 30, 20, 30)$, which amounts to 19600 time series.

4.2   Baselines and Compared Methods

As detailed in the introduction, co-clustering has been the subject of numerous works on various data types, but most of the time on univariate observations matrices (i.e., $d = 1$). To the best of our knowledge, the co-clustering on datasets containing multidimensional cells ($d > 1$) has only been dealt with very recently in the literature, and only a few methods currently exist. Apart from the model-based method from [9], which does not include a dimension reduction aspect and is, therefore, impractical on large datasets, FunLBM [6] is the only existing method dealing with this use case. In addition, we consider two decoupled methods that perform co-clustering by inferring row-partitions and column-partitions independently: a bi-dimensional Gaussian Mixture Model (BGMM) and a bi-dimensional DPMM (BDPMM). This benchmark compares the block partitions quality, but also the results of the associated model selection step, described by the selected number of blocks, compared to the true number (5×5). For BGMM and FunLBM, which are parametric approaches, this selection step is performed by a grid-search with the ICL selection [3, 26] and with a maximum of seven row clusters and seven column clusters. In the non-parametric cases, this selection is natively performed with the inference, with hyper-parameters $\alpha = 0.1, \beta = 0.1$, and G_0 specified as described in Sect. 3.4. For each methods, the results are the performances mean on 10 runs - or the median in the case of the number of clusters. These performances are shown in Table 1.

FunNPLBM has the upper-hand in this benchmark, and estimates the correct structure, while its parametric counterpart FunLBM slightly underestimates the correct number of blocks. In this setup, FunLBM's performances comes from locally optimal estimations of the candidate models during the grid-search model selection. The same effect explains the performance's gap between BGMM and BDPMM. Overall, the two-steps methods BGMM and BPDMM show the worse results, presumably because they both infer the row and column partitions independently and therefore cannot use the row-clusters informations to help finding the best column-clusters partition, and reciprocally. In conclusion, FunNPLBM is able to simultaneously select the right model and to infer the right partition. On an Ubuntu 18.04 64-bit with Intel® Core™ i7-8850H CPU @ 2.60 GHz × 12 CPU 32 Gib RAM, the computing time of FunNPLBM was $<= 30$ s.

Table 1. Benchmark with Bi-dimensional GMM (BGMM), Bi-dimensional DPMM (BDPMM), functional Latent Block Model (FunLBM), and our proposal FunNPLBM

Score	BGMM	BDPMM	FunLBM	FunNPLBM
ARI	0.558	0.667	0.897	1
RI	0.927	0.958	0.990	1
NMI	0.837	0.905	0.968	1
# Blocks	12	16	22	25

4.3 Hyperparameters Specification Study

The hyperparameter set is composed of the base NIW distribution G_0, the concentration parameters (α, β), the iterations number M, and the preprocessing parameters: the Fourier basis dimension and the number of PCA axes. In an unsupervised context, hyperparameters specification remains an active research topic as there is no label to support hyperparameter inference. Consequently, it is not possible to give definitive and general good specifications choices, which depend on the dataset contents and must be hand-tuned by the experts. These specifications, however, can be based on the knowledge of each hyperparameters impact on funNPLBM's behavior, which we illustrate in the following. In each case, we compare funNPLBM's performances when one hyperparameter varies while keeping the others equal to given default values: $\alpha = \beta = 0.1$, $M = 10$, a Fourier expression basis of dimension 30 and 3 PCA axes.

Concentration Parameters and Number of Iterations. In the funNPLBM setting, as in the DPM, the prior distribution of the number of components is an increasing function of the concentration parameters: without knowledge of the data, the higher the concentration parameters, the higher the probability of producing high numbers of components. Because the whole method is symmetric on the dataset rows and columns, and because the experiment dataset dimensions n and p are equal, we consider the case $\alpha = \beta$. As shown in Fig. 1, the concentration parameters effects are negligible for this dataset and only have an impact when extremely high (>1E12) - in which case the number of components is highly overestimated - or extremely small (<1E–10) - in which case only one-cluster partitions are inferred. This small impact of α comes presumably from the high separation of the components in the time series high-dimensional observation space. This separability is simulated for this experiment but is consistent with what we observe in practice. This separation also explains the small number of iterations needed for convergence (here, less than 4 for $1e{-}6 \leq \alpha \leq 1e10$), and the high stability of the MCMC. For a specific dataset, if α and β's values are complex to specify, we advise, as a simple workaround, to add a Gamma hyperprior assumption for α and β and estimate their values during the inference algorithm, which is a common practice in the DPM setting. It implies, however, to study specification strategies for the Gamma distribution parameters.

Preprocessing Parameters. The Fourier Basis dimension and the PCA axes numbers both influence funNPLBM's performances and affect the trade-off between sparsity and quality of time series representation. Figure 2(a) shows the effects of under-estimating or over-estimating the number of PCA axes. Low numbers of PCA axes (<3) are associated with poor scores and few block components due to poor representations of the time series, which are too close in the projection space. On the contrary, if the number of axes is too high (>= 6), the high dimensionality exaggerates the time series separation, and the component number is overestimated, which explains the sharp decrease of the ARI and

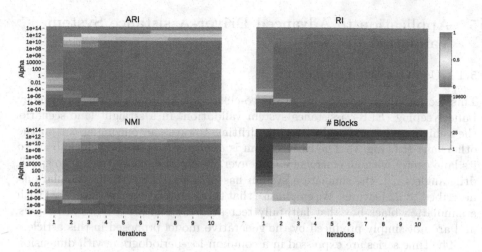

Fig. 1. Scores versus alpha and iterations - with # block in log scale

NMI. In our use cases, we observed that 3 PCA axes lead to the most interesting results. In Fig. 2(b), we observe that, with a fixed number of 3 PCA axes, high polynomial basis dimensions (>50) are correlated with poor scores, presumably because of lower time series representation quality (reflected by the low variance explained score). On the contrary, when this basis dimension is low (<10), the 3-axes PCA adequately represents the information and the variance explained is high (>= 0.97). However, in this case, the Fourier basis dimension is too low to adequately represent the time series, which explains the poor scores. The studied datasets, data generation scripts, and the Scala code used for the benchmark and the experiments are available at the following (anonymized) github repository https://tinyurl.com/4k9jze45, along with the data simulation method. In the next section, a real-case situation is studied and illustrates the method's interest for Advanced Driving-Assistance Systems validation.

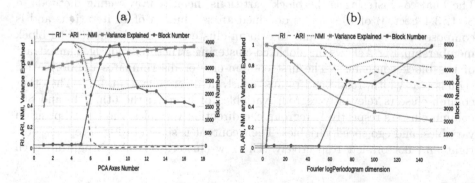

Fig. 2. (a) Scores versus number of PCA axes. (b) Scores versus Fourier logPeriodogram dimension.

5   Application to Advanced Driver-Assistance System Validation

5.1   Use Case Description

This section illustrates the use of the co-clustering approach to the Emergency Lane Keeping (ELK) assistance system validation. In a straight lane scenario, the vehicle under test (called *ego*) is drifting towards an oncoming car on the other lane (c.f. Fig. 3). The ELK system is expected to put the vehicle back to its lane center with an emergency maneuver. With different settings (ego speed, drift angle, ...), the simulation system has produced a set of 400 simulations, described by 22 features. We emphasize that these simulations are generated with a simulation black-box that faithfully recreates the real-life driving conditions, and are not simply produced by the generative model proposed in this article.

The time series are expressed in a common log-periodogram with dimension 40, then reduced to 3 PCA axes. The concentration parameters are set to 1e–2, and the NIW parameters to the default values discussed in Sect. 3.4. The objective is to discriminate simultaneously driving patterns and correlated variables.

Fig. 3. Use case illustration: ego drifts from its lane, crosses the center line and heads toward an oncoming vehicle. The system detects the target and change ego's direction.

5.2   Results

The final co-clustering is the block partitions mean (c.f. averaging methods in Sect. 3.4–Sect. 3) of 10 samples obtained after a burnin of 10 iterations and is composed of 6 row-clusters and 13 column-clusters. With color indicating block membership, Fig. 4 shows the global co-clustering structure, and Fig. 5 an extract of the block contents.   The first column-cluster discriminates uninformative signals (car width, road bend radius, constant inactive system, ...). The other column-clusters relevantly regroup variables of interest: the 6th, 7th, and 8th column clusters respectively regroup ego direction variables, ego lateral position variables, and ego speed variables. Their content is shown in Fig. 5 top-left, top-right and bottom-left respectively. The row-clustering is also insightful: each

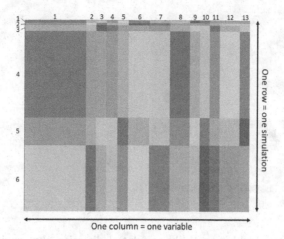

One row = one simulation

One column = one variable

Fig. 4. Resulting co-clustering on Emergency Lane Keeping (ELK) dataset. The result consists of 6 row-clusters and 13 column-clusters

row-cluster correspond to well-separated driving behaviors. This separation is best seen in Fig. 5 (top-right) that shows the following driving behaviors: 1) ego drifting left and the ELK system failing (light green); 2) the symmetric behavior on the right (dark green); 3) the ELK system correcting the car trajectory (light orange).

Finally, the three other row-clusters (regrouping the remaining 5% of the observations) are composed of outliers simulations, with driving behavior displayed on Fig. 5 (bottom-right). In this situation, the oncoming car is correctly detected, and ego heading is changed accordingly, but not enough to prevent the collision. In conclusion, funNPLBM has correctly discriminated uninformative signals while creating meaningful clusters of features and clusters of simulations. From this information, it is easy to visualize the variety of driving behaviors that compose our datasets and understand them from the variable perspectives, which was the original objective of the application. The next step is to link the driving behavior to the control logic parameters and, if need be, refine them to reach the performance objectives. With the same computer specifications than for the experiments (c.f. Sect. 4.2), the computing time was <20 s.

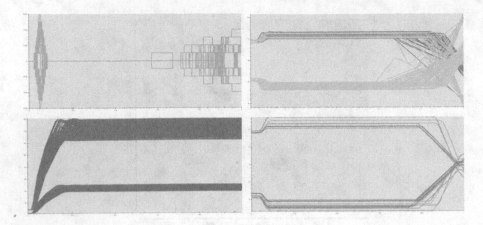

Fig. 5. Top-left: two highly negatively correlated direction change signals; top-right: ego lateral position in the 3 biggest observation clusters; bottom-left: 2 correlated speed variables; bottom-right: 3 outlier driving pattern in the 3 smallest observation clusters.

6   Conclusion and Future Work

This paper describes FunNPLBM, a Bayesian non-parametric based method that addresses the problem of co-clustering multivariate time series. This work proposes the first Bayesian non-parametric co-clustering method dedicated to functional data analysis, the description of an adapted collapsed Gibbs sampling, and a more compact definition of the NPLBM model.

This method regroups redundant features, discriminates uninformative ones, and provides the user with a two-dimensional analysis of a multivariate time series dataset. The hyperparameters (concentrations parameters, components parameters) specifications are discussed and experimented on a simulated dataset, and a benchmark is presented that shows FunNPLBM adequacy in a context that matches our assumptions. Finally, the method is applied to a real-case dataset from the autonomous driving system validation domain. In this application, FunNPLBM proves its ability to create meaningful clusters of driving behavior and correlated variables simultaneously.

We are confident that FunNPLBM can be useful in other domains dealing with correlated temporal variables. For instance in industrial contexts for sensor anomaly detection or predictive maintenance, in health for ECG and biological signals data analysis, in finance for stock trade data analysis. In addition, The method can also be applied to anomaly detection thanks to the native production of probabilistic predictive intervals and supervised classification by simply constraining the row and column partitions values. We consider two extensions: a) higher-order tensor co-clustering; b) relaxing the model to multi-clustering. These perspectives will be addressed in future work.

References

1. Aghabozorgi, S., Shirkhorshidi, A.S., Wah, T.Y.: Time-series clustering-a decade review. Inform. Syst. **53**, 16–38 (2015)
2. Bagnall, A., Lines, J., Bostrom, A., Large, J., Keogh, E.: The great time series classification bake off: a review and experimental evaluation of recent algorithmic advances. Data Min. Knowl. Discov. **31**(3), 606–660 (2016). https://doi.org/10.1007/s10618-016-0483-9
3. Biernacki, C., Celeux, G., Govaert, G.: Assessing a mixture model for clustering with the integrated completed likelihood. IEEE Trans. Patt. Anal. Mach. Intell. **22**(7), 719–725 (2000)
4. Blei, D.M., Jordan, M.I., et al.: Variational inference for dirichlet process mixtures. Bayesian Anal. **1**(1), 121–143 (2006)
5. Boullé, M.: Functional data clustering via piecewise constant nonparametric density estimation. Patt. Recogn. **45**(12), 4389–4401 (2012)
6. Bouveyron, C., Bozzi, L., Jacques, J., Jollois, F.X.: The functional latent block model for the co-clustering of electricity consumption curves. J. R. Stat. Soc. Ser. C (Appl. Stat.) **67**(4), 897–915 (2018)
7. Caiado, J., Crato, N., Peña, D.: Comparison of times series with unequal length in the frequency domain. Commun. Stat. Simul. Comput. **38**(3), 527–540 (2009)
8. Celeux, G., Frühwirth-Schnatter, S., Robert, C.P.: Model selection for mixture models—perspectives and strategies. In: Handbook of Mixture Analysis (2018)
9. Chamroukhi, F., Biernacki, C.: Model-based co-clustering of multivariate functional data. In: Proceedings of the 61st World Statistics Congress (2017)
10. Côme, E., Jouvin, N., Latouche, P., Bouveyron, C.: Hierarchical clustering with discrete latent variable models and the integrated classification likelihood. In: Advances in Data Analysis and Classification, pp. 1–30 (2021)
11. Dempster, A.P., Laird, N.M., Rubin, D.B.: Maximum likelihood from incomplete data via the EM algorithm. J. R. Stat. Soc. Ser. B (Methodological) **39**(1), 1–22 (1977)
12. Dhillon, I.S.: Co-clustering documents and words using bipartite spectral graph partitioning. In: Proceedings of the seventh ACM SIGKDD International Conference on Knowledge Discovery and Data Mining, pp. 269–274 (2001)
13. Dhillon, I.S., Mallela, S., Modha, D.S.: Information-theoretic co-clustering. In: Proceedings of the Ninth ACM SIGKDD International Conference on Knowledge Discovery and Data Mining, pp. 89–98 (2003)
14. Escobar, M.D.: Estimating normal means with a dirichlet process prior. J. Am. Stat. Assoc. **89**(425), 268–277 (1994)
15. Forest, F., Mourer, A., Lebbah, M., Azzag, H., Lacaille, J.: An invariance-guided stability criterion for time series clustering validation. In: International Conference on Pattern Recognition (ICPR) (2020)
16. Gelman, A., Carlin, J.B., Stern, H.S., Dunson, D.B., Vehtari, A., Rubin, D.B.: Bayesian Data Analysis. CRC Press, Boca Raton (2013)
17. Glassen, T.J., von Oertzen, T., Konovalov, D.A.: Finding the mean in a partition distribution. BMC Bioinform. **19**(1), 1–10 (2018)
18. Goffinet, E., Coutant, A., Lebbah, M., Azzag, H., Giraldi, L.: Conditional latent block model: a multivariate time series clustering approach for autonomous driving validation. arXiv preprint arXiv:2008.00946 (2020)
19. Govaert, G., Nadif, M.: Clustering with block mixture models. Patt. Recogn. **36**(2), 463–473 (2003)

20. Govaert, G., Nadif, M.: Block clustering with bernoulli mixture models: comparison of different approaches. Comput. Stat. Data Anal. **52**(6), 3233–3245 (2008)
21. Govaert, G., Nadif, M.: Co-clustering: Models, Algorithms and Applications. John Wiley & Sons, Hoboken (2013)
22. Görür, D.: Nonparametric bayesian discrete latent variable models for unsupervised learning. Doctoral thesis, Technische Universität Berlin, Fakultät IV - Elektrotechnik und Informatik, Berlin (2007)
23. Hastie, D.I., Liverani, S., Richardson, S.: Sampling from dirichlet process mixture models with unknown concentration parameter: mixing issues in large data implementations. Stat. Comput. **25**(5), 1023–1037 (2015)
24. Huelsenbeck, J.P., Andolfatto, P.: Inference of population structure under a dirichlet process model. Genetics **175**(4), 1787–1802 (2007)
25. Jacques, J., Biernacki, C.: Model-based co-clustering for ordinal data. Comput. Stat. Data Anal. **123**, 101–115 (2018)
26. Keribin, C., Brault, V., Celeux, G., Govaert, G.: Estimation and selection for the latent block model on categorical data. Stat. Comput. **25**(6), 1201–1216 (2014). https://doi.org/10.1007/s11222-014-9472-2
27. Kluger, Y., Basri, R., Chang, J.T., Gerstein, M.: Spectral biclustering of microarray data: coclustering genes and conditions. Genome Res. **13**(4), 703–716 (2003)
28. Konovalov, D.A., Litow, B., Bajema, N.: Partition-distance via the assignment problem. Bioinformatics **21**(10), 2463–2468 (2005)
29. Křivánek, M., Morávek, J.: Np-hard problems in hierarchical-tree clustering. Acta Inform. **23**(3), 311–323 (1986)
30. Laclau, C., Redko, I., Matei, B., Bennani, Y., Brault, V.: Co-clustering through optimal transport. In: International Conference on Machine Learning. PMLR (2017)
31. Long, B., Zhang, Z., Yu, P.S.: Co-clustering by block value decomposition. In: Proceedings of the Eleventh ACM SIGKDD International Conference on Knowledge Discovery in Data Mining, pp. 635–640 (2005)
32. Meeds, E., Roweis, S.: Nonparametric Bayesian Biclustering. Tech. rep, Citeseer (2007)
33. Meguelati, K., Fontez, B., Hilgert, N., Masseglia, F.: Dirichlet process mixture models made scalable and effective by means of massive distribution. In: Proceedings of the 34th ACM/SIGAPP Symposium on Applied Computing, pp. 502–509 (2019)
34. Neal, R.M.: Markov chain sampling methods for dirichlet process mixture models. J. Comput. Graph. Stat. **9**(2), 249–265 (2000)
35. Nguyen, V.A., Boyd-Graber, J., Altschul, S.: Dirichlet mixtures, the dirichlet process, and the structure of protein space. J. Comput. Biol. **20**, 1—18 (2013)
36. Ramsay, J., Silverman, B.: Principal components analysis for functional data. In: Functional Data Analysis. Springer Series in Statistics, pp. 147–172. Springer, New York (2005). https://doi.org/10.1007/0-387-22751-2_8
37. Ross, G.J., Markwick, D.: Dirichlet process: an r package for fitting complex Bayesian nonparametric models (2018)
38. Schlüter, K., Drenckhahn, D.: Co-clustering of denatured hemoglobin with band 3: its role in binding of autoantibodies against band 3 to abnormal and aged erythrocytes. Proc. Natl. Acad. Sci. **83**(16), 6137–6141 (1986)
39. Schuurman, N., Grasman, R., Hamaker, E.: A comparison of inverse-Wishart prior specifications for covariance matrices in multilevel autoregressive models. Multivar. Behav. Res. **51**(2–3), 185–206 (2016)

40. Sethuraman, J.: A constructive definition of dirichlet priors. Stat. Sin. **4**, 639–650 (1994)
41. Slimen, Y.B., Allio, S., Jacques, J.: Model-based co-clustering for functional data. Neurocomputing **291**, 97–108 (2018)
42. Williamson, S., Dubey, A., Xing, E.: Parallel markov chain monte carlo for nonparametric mixture models. In: International Conference on Machine Learning (2013)
43. Xanthopoulos, P.: A review on consensus clustering methods. In: Optimization in Science and Engineering, pp. 553–566. Springer, New York (2014). https://doi.org/10.1007/978-1-4939-0808-0
44. Xu, D., et al.: Deep co-clustering. In: Proceedings of the 2019 SIAM International Conference on Data Mining, pp. 414–422. SIAM (2019)
45. Yan, Y., Chen, L., Tjhi, W.C.: Fuzzy semi-supervised co-clustering for text documents. Fuzzy Sets Syst. **215**, 74–89 (2013)

TRAMESINO: Traffic Memory System for Intelligent Optimization of Road Traffic Control

Cristian Axenie[1], Rongye Shi[2(✉)], Daniele Foroni[1], Alexander Wieder[1], Mohamad Al Hajj Hassan[1], Paolo Sottovia[1], Margherita Grossi[1], Stefano Bortoli[1], and Götz Brasche[1]

[1] Intelligent Cloud Technologies Lab, Huawei Munich Research Center, Riesstrasse 25, 80992 Munich, Germany
cristian.axenie@huawei.com
[2] EI Intelligence Twins Program, Huawei Cloud BU, Shenzhen, China
shirongye@huawei.com

Abstract. Whether efficient road traffic control needs accurate modelling is still an open question. Additionally, whether complex models can dynamically adapt to traffic uncertainty is still a design challenge when optimizing traffic plans. What is certain is that the highly non-linear and unpredictable real-world road traffic situations need timely actions. This study introduces TRAMESINO (TRAffic Memory System INtelligent Optimization). This novel approach to traffic control models only relevant causal action-consequence pairs within traffic data (e.g. green time - car count) in order to store traffic patterns and retrieve plausible decisions. Multiple such patterns are then combined to fully describe the traffic context over a road network and recalled whenever a new, but similar, traffic context is encountered. The system acts as a memory, encoding and manipulating traffic data using high-dimensional vectors using a spiking neural network learning substrate. This allows the system to learn temporal regularities in traffic data and adapt to abrupt changes, while keeping computation efficient and fast. We evaluated the performance of TRAMESINO on real-world data against relevant state-of-the-art approaches in terms of traffic metrics, robustness, and runtime. Our results emphasize TRAMESINO's advantages in modelling traffic, adapting to disruptions, and timely optimizing traffic plans.

1 Introduction

Solving traffic congestion in urban agglomerations is still a problem resistant to straightforward solutions despite the large amount of research and systems developed to analyze [20], model [23], and control road traffic [26]. Systems

---

C. Axenie, R. Shi, D. Foroni, A. Wieder, M. A. H. Hassan, P. Sottovia, M. Grossi and S. Bortoli—Authors contributed equally to this research.

V. Lemaire et al. (Eds.): AALTD 2021, LNAI 13114, pp. 88–103, 2021.
https://doi.org/10.1007/978-3-030-91445-5_6

deployed in real-world [6, 10, 14] use a traffic model [7, 24] that heavily influences the run-time performance of the overall system. Basically, the role of the traffic model is to describe the dynamics of the traffic flow and to cope, eventually, with unforeseen deviations (i.e. disruptions) in traffic patterns [24]. But, in order to achieve that, the system needs to optimize multiple metrics, such as spatial and temporal traffic demand, traffic volume [3]. This implies a substantial computational cost that might hinder the overall real-time capabilities of the system and increase the cost of large scale traffic optimization. It is the system designer's duty to make a trade-off between two dimensions, namely performance and execution time. The present study addresses the problem of such costly optimization routines and explores a novel approach to speed-up traffic control, named TRAMESINO (TRAffic Memory System INtelligent Optimization). At the core of TRAMESINO is the capability to exploit the similarity and invariant features of traffic flow patterns. Basically, by storing relevant causal patterns (i.e. action/consequence: allocated green time/measured car count) in traffic flows, one can bypass the costly constrained optimization routines typically employed in traffic control systems. Such patterns can be correlated in time to fully describe the traffic context in an entire region. Given "cues" of traffic data (i.e. current car count), the system can "recover" a plausible cause (i.e. the allocated green time). To achieve this TRAMESINO uses:

- an efficient encoding scheme for traffic timeseries covariates;
- a mechanism storing associations among traffic timeseries covariates;
- an efficient learning framework to natively process the encoded quantities and implement the association dynamics.

In the remainder of this section, we ground our contribution and emphasize those relevant features and drawbacks of adaptive traffic control systems motivating our study.

1.1 Optimization-Based Adaptive Traffic Control Systems

Traditionally, flow optimization for coordinated traffic signals is based on average travel times between intersections and average traffic volumes at each intersection [9]. However, most of these approaches do not consider the stochastic nature of high-resolution field traffic data or capture it through computationally expensive processes, such as Markov Decision Processes (MDP) [22]. Beyond stochasticity, the community also explored the use of mixed-integer linear programming (MILP) for optimizing the control of traffic signals, in particular, offsets, split times, and phase orders [11]. The approach provided optimal results but with a high computational cost. Additionally, such systems couldn't handle changes of the controlled variables in real-time due to the optimization process that needs to iterate to convergence. In a first attempt to exploit the periodic nature of the traffic signals, the work in [18] formulated the traffic light optimization into a continuous optimization problem without integer variables, by modeling traffic flow as sinusoidal. The system solved a convex relaxation of the non-convex problem using a tree decomposition reduction with very good performance in simulations, but it lacked the capability to scale and adapt to traffic disruptions. Finally, relying on predicting arrivals at coordinated signal approaches the work in [2] proposed the link pivot algorithm that assumed nearest-neighbor interactions between signals in cyclic flow profiles to model traffic flows. Despite its well performing optimization, the algorithm couldn't handle unpredictable changes in platoon shapes (i.e. occasionally caused by platoon splitting and merging) or prediction during saturated conditions (i.e. traffic jams, accidents) limiting its use in real-world deployments. As we briefly emphasized hitherto, aspects such as stochasticity, simultaneous traffic assignment and traffic signal calculation, periodicity, regional scaling, and real-time constraints, describe real-world traffic situations. Each system excels in handling a sub-set of these aspects only and cannot capture their combined impact on traffic dynamics.

1.2   Beyond Optimization

Historically, neural networks were employed in traffic control to exploit its intrinsic temporal dynamics. A reference work in this category is the study of [16] which proposed a Hopfield network-based system designed to capture temporal patterns. Opposite to optimization approaches, such a system exploited the interaction between neurons whose dynamics modelled traffic signals state changes and stochasticity. These first steps away from optimization, were extended in [17] by emphasizing the purpose of feedback loops for decreasing the differences of the conflicting flows, measured during a congestion or large number of waiting vehicles. This solution enabled regional scaling and simultaneous traffic assignment and traffic signal calculation using the same network, by exploiting the capability to describe and solve a constraint satisfaction problem of Hopfield networks [8]. Using a simplified, linear Hopfield neural network, the study in [12] proposed a system capable of solving an arbitrary set of (linear) equations through online learning. Interestingly, the typical Hopfield network was augmented with an additional feed-forward layer used to compute the Moore-Penrose Generalized Inverse (i.e. pseudoinverse) of the weight connection matrix. Calculating the pseudoinverse, allowed the system to actually compute a "best fit" (in least squares sense) solution to the evolving traffic dynamics model (i.e. unexpected disruptions "move" the optimum in solution space) in a parallel fashion. Addressing the scaling problem, the study in [27] employed an augmented Hopfield network to solve mixed integer programming. The approach exploited the temporal dynamics of the Hopfield network to find better solutions than Lagrangian relaxation and only very rarely converged to unfeasible solutions. Finally, despite the advantages that systems, such as Hopfield networks, have, there are some known barriers to deploy them to real-world scenarios. A first aspect refers to the limited memory capacity of Hopfield networks, and the actual computational cost of storing a large number of memories - which increases with the number of neurons. Another strong limitation is the pattern orthogonality assumption, which limits the recall accuracy, especially at scale. Considering the unique and optimal recall of Hopfield networks, there are strong limitations due to the existence of local minima and spurious states of attraction. Such a limitation is stronger in the case of storing high-dimensional traffic contexts, where due to increased similarity, identifying the discrepancy is difficult, especially in the presence of a large number of states of attraction.

1.3 Motivation and Contributions

Besides the relevant aspects already mentioned, a significant drawback of existing traffic control systems is that they fail to fully exploit the causal coupling (or associations) between traffic control signals and traffic flow dynamics. It is known that, despite being highly nonlinear, traffic dynamics is regular on certain timescales. Such regularities together with available sensory data can be used to judiciously extract traffic contexts that can be subsequently used in optimizing future traffic situations. Basically, the associations among the control signals (i.e. green time/red time) and the measured outcomes (i.e. flow of cars) capture the dynamics of traffic on a road, intersection, or region. Obviously, in order to optimize flow and minimize delay time, the traffic control system would need to find the best traffic light timing. This functionality is described in Fig. 1.

In this context, our contribution focuses on four main points:

- optimizing the time for decision-making and "short-circuit" re-computation of a control signal (i.e. green time allocation) by exploiting previously learnt patterns of traffic context (i.e. traffic flow – green time pair). Metaphorically, TRAMESINO accumulates wisdom over traffic optimization, and uses the acquired knowledge to bypass possibly computationally complex decision-making processes based solely on the ongoing traffic perception.
- representing traffic contexts (i.e. regional traffic flow, local allocated green times, etc.) as a "memory", basically a high-dimensional numeric vector depicting the traffic state at a certain moment in time. Additionally, such memories can be stored and recovered using a learning system, which is at the core of TRAMESINO. This way TRAMESINO can exploit the descriptive power of pairs of actions and their outcomes in order to learn memories from historical data. Such memories of associations speed-up operation, when facing new traffic situations by recalling the most similar (previously seen) traffic context. To support this speed-up, the contexts are represented using high-dimensional vectors, which map the complex dynamics to simple (algebraic) operations in high-dimensional spaces.
- exploiting learned context associations among patterns of traffic (i.e. traffic flow – allocated green time) to infer what would be the most plausible traffic flow when a control signal (i.e. green time) is available and what would be the green time for measured flow values. In other words, given a partial context, TRAMESINO recalls the most similar context learnt in the past by restoring the missing part of the context (i.e. either green time for measured flow or flow for applied green time) - similarly to an autoassociative memory.
- release of a new real-world dataset, used in the TRAMESINO experiments, which contains 74 days of real urban road traffic data from 8 crosses in a city in China.

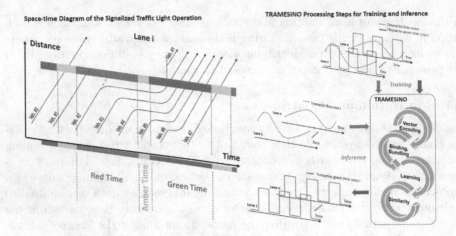

Fig. 1. TRAMESINO system functionality overview.

The main problems the proposed system solves are:

- efficiently representing traffic context using the measured data (i.e. traffic flow) and control signals (i.e. allocated green time) in a system capable of learning multiple associations among such causal data encoded in efficient high-dimensional vectors;
- avoiding costly optimization methods and control signal re-computation by exploiting previously learnt patterns of traffic data and infer, given partial information (i.e. either traffic flow or allocated green time), what would be the best corresponding full context corresponding to the partial information;
- scalability through storing multiple memories (i.e. multiple full traffic contexts) and deployment at different granularity (e.g. per lane, per direction, per intersection);
- the efficient computation of traffic control signals (i.e. green time) that embed and exploit the intrinsic traffic constraints and physics without an explicit need to model the constraints;

2   Materials and Methods

In this study, we introduce TRAMESINO, a flexible framework and system capable to learn and store associations among measured traffic data (e.g. traffic flow) and the corresponding traffic control signal generating it (i.e. allocated green time). In order to speed up computation in similar, but novel, traffic situations, the system recalls the most plausible learnt association.

2.1   Introducing TRAMESINO

TRAMESINO is an associative memory system for traffic flow optimization. The system builds a vector description of the current traffic context from timeseries

of specific traffic data (i.e. flow of cars, green time, traffic density). The key
ingredient of TRAMESINO is the Holographic Reduced Representation (HRR)
[19], responsible for the traffic data encoding, learning, and computation with
the encoded quantities. Such a framework demonstrated already that structured
vector-representations are able to capture relations and mutual influence between
multiple traffic context data [15].

HRR are a type of Vector Symbolic Architectures (VSAs) [5] that describe
a family of modelling approaches to represent physical quantities by mapping
them to (high-dimensional) vectors. Beside the numerical structure underlying
the vectors, the core computational components of a VSA are a measure of similarity and typically two algebraic operations, namely superposition and binding.
Superposition implements the basic addition and combines multiple vectors to
create a vector similar to the input vectors. Binding implements the basic multiplication in order to produce highly dissimilar response to both input vectors.
A very important aspect is the fact that binding is invertible and preserves distance metrics which support the associative memory implementation. Within
TRAMESINO, superposition allows storing multiple traffic contexts defined by
available traffic data (i.e. flow, green time, cycle time, phase length), whereas
binding provides the core mechanism to recall previously stored contexts given a
similarity metric. An important property of the high-dimensional vector space in
TRAMESINO is that with a very high probability all stored vectors are dissimilar to each other (i.e. quasi-orthogonal). This enables the system to implement
the associative memory behavior using simple operations in high dimensions.
Finally, unlike many traditional neural networks, HRR do not rely on backpropagation but rather on algebraic operations on high-dimensional vectors which are
embarrassingly parallel operations that can be performed efficiently (in principle,
in constant time).

Data Representation. TRAMESINO uses high-dimensional HRR vectors
and operations to represent traffic contexts (i.e. action-consequence pairs)
and do computation (i.e. associative memory). Intuitively, for practical use,
TRAMESINO needs to store multiple such contexts as memories to be able
to handle arbitrary new contexts. As mentioned, in order to store multiple vectors encoding traffic contexts, TRAMESINO utilizes bundling, which accounts
for an element-wise addition of the vectors. For the recall phase, TRAMESINO
utilizes binding, which is basically implementing a circular convolution. We now
introduce the formalism behind the specific HRR operations in TRAMESINO.

HRR allow for complex vector values, i.e., $N \subseteq \mathbb{C}$ and use a multiplication
operation $\circledast$ based on circular convolution. For any two vectors of size D, $\mathbf{x}, \mathbf{y} \in
V_D(N)$, circular convolution $\circledast$ is defined as

$$\mathbf{z} = \mathbf{x} \circledast \mathbf{y} \qquad \text{with } z_j = \sum_{k=0}^{D-1} x_k y_{(j-k) mod D}. \tag{1}$$

Circular convolution can efficiently be computed using the Discrete Fourier Transform (DFT) [1] defined as the function

$$DFT : \mathbf{C}^D \to \mathbf{C}^D, \mathbf{x} \to \left(\sum_{j=0}^{D-1} x_j \zeta_D^{-jk} \right)_{k=0}^{D-1} \qquad \text{with } \zeta_D = \exp\left(\frac{i2\pi}{D}\right). \quad (2)$$

Similarly, the Inverse Discrete Fourier Transform (IDFT) is defined as the function

$$IDFT : \mathbf{C}^D \to \mathbf{C}^D, \mathbf{x} \to \left(\frac{1}{D} \sum_{j=0}^{D-1} x_j \zeta_D^{jk} \right)_{k=0}^{D-1}. \quad (3)$$

In TRAMESINO, we make use of the fact, that circular convolution can be written as a combination of the DFT, IDFT, and element-wise multiplication $\odot$ [19]. Using the convolution theorem, we can calculate the circular convolution of any two vectors $\mathbf{v}, \mathbf{w} \in V_D(N)$ by

$$\mathbf{v} \circledast \mathbf{w} = IDFT \left(DFT(\mathbf{v}) \odot DFT(\mathbf{w}) \right), \quad (4)$$

with $\odot$ denoting element-wise multiplication in this case. This induces that circular convolution obeys the same commutative and associative rules as element-wise multiplication. Additionally, we define the convolutive power as the real part of the transformed vector

$$\mathbf{v}^p := \Re \left(IDFT \left(DFT \left(\mathbf{v} \right)^p \right) \right), \quad (5)$$

This operation is used when recalling a traffic memory. This involves building the HRR vector of a partial context (i.e. traffic flow car count), bundling and binding it to existing memories, and then computing the similarity.

Next, we describe the encoding, its constraints, and the considerations to handle temporal aspects (i.e. traffic contexts are timeseries of various traffic measured quantities). TRAMESINO uses the unitary base of vectors $\mathbf{b}$ for encoding (real-valued) scalar traffic quantities (i.e. flow of cars, green time) in high-dimensional HRR vectors, which are in fact combinations of basis vectors using simple algebraic operations. Additionally, it uses $C_0, \dots, C_D$ to represent each type of traffic data and $T_0, \dots, T_D$ for encoding the temporal structure (i.e. timestamps). As a design choice, we use unitary vectors u, since they have some desirable properties, namely $|\mathbf{u}| = 1$, $\mathbf{u}^p$ is still unitary for any $p \in \mathbb{R}$, and convolution with unitary vectors preserves the norm, i.e., $|\mathbf{v}| = |\mathbf{v} \otimes \mathbf{u}|$ for any other vector $\mathbf{v}$. We can now create actual HRR vectors V_i of different traffic quantities values v_i as

$$V_i = \sum_{j=1}^{D} C_j \otimes \mathbf{b}^{v_j \cdot s}, \quad (6)$$

where s is a scaling factor. To additionally encode the temporal structure, we simply bind each traffic quantity vector V_i to a vector T encoding the timestamp,

$$V_T = \sum_{i=1}^{D} \left(\sum_{j=1}^{D} C_j \otimes \mathbf{b}^{v_j \cdot s} \right) T_i. \quad (7)$$

Learning and Inference. In TRAMESINO, the HRR traffic data representation and the HRR binding and bundling operations, are implemented in the
Neural Engineering Framework (NEF) [4]. NEF offers a systematic method of
"compiling" high-level descriptions, such as vector convolution, correlation, and
similarity, into synaptic connection weights between populations of spiking neurons with efficient learning capabilities. In NEF, neural populations represent
time-varying signals, such as traffic flow data, through their spiking activity.
Such signals drive neural populations based on each neuron's tuning curve, which
describes how much a particular neuron will fire as a function of the input signal.

Formally, we consider A a population of $N \in \mathbb{N}$ neurons encoding a subset
V of a real-valued vector space, i.e., $V \subseteq \mathbb{R}^n$, representing measurable traffic
quantities. Given a function $\mathbf{x(t)} : \mathbb{R} \to V$, we can write the activity a_i at time
t of the i-th neuron in a neural population encoding a time-varying vector (e.g.
traffic flow data) $\mathbf{x}(t)$ as a spike train,

$$a_i\left(\mathbf{x}(t)\right) = \sum_{j=1}^{m_i} \delta(t - t_j) = G_i(\alpha_i \langle \mathbf{e}_i, \mathbf{x}(t) \rangle + J_i) \quad \text{for } 1 \leq i \leq N, \quad (8)$$

where G_i is the neural non-linearity, α_i is the gain of the neuron, $\mathbf{e}_i$ is the
neuron's preferred encoding vector, J_i describes the neural background activity,
and t_j are the m_i spike-times of the i-th neuron, and $\langle . \rangle$ is the inner product. To
decode the traffic quantities $\mathbf{x}(t)$ back out of the neural population A, the spike
train is convolved $*$ with an exponentially decaying filter $h : \mathbb{R} \to \mathbb{R}$ resulting in

$$\tilde{a}_i\left(\mathbf{x}(t)\right) = \sum_{j=1}^{m_i} h(t) * \delta(t - t_j) = \sum_{j=1}^{m_i} h(t - t_j). \quad (9)$$

We consider here the exponential decaying filter given by $h : \mathbb{R} \to \mathbb{R}, t \to e^{\frac{-t}{\tau_p}}$,
where τ_p is the post-synaptic time constant. Through filtering we obtain an
estimation $\hat{\mathbf{x}}(t)$ of the original input $\mathbf{x}(t)$ as a weighted sum with some decoder
values $\mathbf{d}_i$

$$\hat{\mathbf{x}}(t) = \sum_{i=1}^{N} \tilde{a}_i\left(\mathbf{x}(t)\right) \mathbf{d}_i. \quad (10)$$

To calculate the optimal decoders $\mathbf{d}_i$, the system needs to minimize the error
between input $\mathbf{x}(t)$ and decoded output $\hat{\mathbf{x}}(t)$

$$E = \int \left(\mathbf{x}(t) - \sum_{i=1}^{N} \tilde{a}_i\left(\mathbf{x}(t)\right) \mathbf{d}_i\right)^2 d\mathbf{x}(t). \quad (11)$$

NEF solves for the decoders $\mathbf{d}_i$ by default using an efficient least squares optimization [4].

Encoding and decoding operations on NEF neural populations representations allow us to encode traffic flow signals over time, and decode transformations (i.e. mathematical functions) of those signals. In fact, NEF allows us to

decode arbitrary transformations of the input traffic data by computing functions across the connections between the populations of neurons encoding the traffic data. For instance, if we consider A resp. B populations of N resp. M neurons encoding a time-varying vector $\mathbf{x}(t) \in V \subset \mathbb{R}^n$ (e.g. traffic flow) resp. $\mathbf{y}(t) \in W \subset \mathbf{R}^m$ (e.g. traffic density) and a function $f : V \to W \subset \mathbf{R}^m$. In order to approximate the function f (i.e. traffic flow - density dependency) across a connection from population A to population B, TRAMESINO calculates a set of decoder values $\mathbf{d}_i^f$ for population A by minimizing the error

$$E = \int \left(f(\mathbf{x}(t)) - \sum_{i=1}^{N} \tilde{a}_i \left(\mathbf{x}(t) \right) \mathbf{d}_i^f \right)^2 d\mathbf{x}(t). \tag{12}$$

Given encoders $\mathbf{e}_j^B$ and gain α_j^B for $1 \leq j \leq M$ of population B, we can derive a weight matrix for the connection from A to B approximating the function f by

$$w_{ij} = \alpha_j^B \mathbf{d}_i^f L \mathbf{e}_j^B \quad \text{for } 1 \leq i \leq N \text{ and } 1 \leq j \leq M, \tag{13}$$

where L is a $M \times N$ linear operator. Here, NEF makes the assumption, that connection weights can be factored into encoders, decoders, and a transform. Finally, in order to implement the associative memory behavior, we need to describe the dynamics of such an operation. But first we introduce how can we implement such dynamics in populations of spiking neurons. If we consider A a population of neurons with an incoming connection approximating the function $f : V \to W \subset \mathbb{R}^m$ and a recurrent connection approximating the function $g : W \to W$ (cf. Fig. 2). Thus, the overall function the population is approximating is

$$\mathbf{y}(t) = h(t) * (f(\mathbf{x}(t)) + g(\mathbf{y}(t))) \tag{14}$$

with exponential decaying filter function $h : \mathbb{R} \to \mathbb{R}, t \to e^{\frac{-t}{\tau}}$. By setting the functions $g(\mathbf{y}(t)) = \tau a(\mathbf{y}(t)) + \mathbf{y}(t)$ and $f(\mathbf{x}(t)) = \tau b(\mathbf{x}(t))$ - with a and b arbitrary nonlinear functions - we obtain a neural model approximating a dynamical system. The learning rule implementing the autoassociative memory needs to modify the encoding vectors of active neurons to be selective to an input vector (i.e. a partial context, traffic flow). Basically, this operation adjusts the connection weights so that a small number of distinct neurons respond to each such partial traffic context - by triggering the memory most similar to it. For this

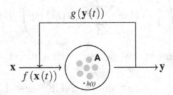

Fig. 2. Dynamics implementation of TRAMESINO associative memory.

we used the three layer neural autoassociator using NEF spiking neurons from
[25]. Given a traffic context vector x encoded by the activity of the input neural population, the filtered activity $a(t)$ of neurons in the middle layer, and the
matrix $\mathbf{e}$ whose rows are the "preferred traffic context" vectors of the middle
layer neurons, we modify the "preferred traffic context" vectors of the middle
layer neurons according to:

$$\frac{\partial \mathbf{e}(t)}{\partial t} = -\eta a(t)\mathbf{e}(t) + \eta(a(t)\mathbf{x}^T(t), \tag{15}$$

where η is the learning rate. Changing the "preferred traffic context" vectors
corresponds to changing the connection weights using a local learning rule in
Eq. 15. This system has been proven to have high accuracy, a fast, feed-forward
recall process, and efficient scaling, requiring a number of neurons linear in the
number of stored associations.

Parametrization. In all our experiments, within TRAMESINO[1], traffic flow
readings from an urban region were concatenated in a context vector (i.e. memory) of size $D = 1024$, each encoding neural population had a size of 100 neurons,
a new memory was stored for each traffic light and each phase every $n = 10$ traffic light cycles (i.e. accounting for a memory every approx. 5 min), and a new
green time was recalled at the generation of each new plan (i.e. approx. every
2 min). Note that, increasing the number of neurons (i.e. ≥ 1000) provides a more
accurate encoding and, hence results, but the computation time increase supra-
linearly. Our parametrization reflects the trade-off to make for the computation
time gain. Figure 3 provides an overview on the traffic context data, the encoding
process, and the similarity calculation processes, respectively.

3   Experiments and Results

In our experiments, we used the SPRING-MUSTARD (Spring season Multi-
cross Urban Signalized Traffic Aggregated Region Dataset) real-world dataset,
which contains 74 days of real urban road traffic data from 8 crosses in a city
in China[2]. The road network layout is depicted in Fig. 4a. In order to perform
experiments and evaluate the system, we simulated the real-world traffic flows in
the Simulator for Urban Mobility (SUMO) [13]. The realistic vehicular simulator
generates routes, vehicles, and traffic light signals that reproduce the real car
flows in the real-world dataset. In order to evaluate the adaptation capabilities,
we systematically introduced progressive flow magnitude disruptions over the
74 days of traffic flow data. Such degenerated traffic conditions describe non-
recurrent events such as sport events, accidents or adverse weather, for instance.
More precisely, accidents and adverse weather typically determine a decrease in

---

[1] Codebase at: https://github.com/omlstreaming/aaltd2021.
[2] The SPRING-MUSTARD real-world dataset used in our experiments is available at:
http://doi.org/10.5281/zenodo.5025264.

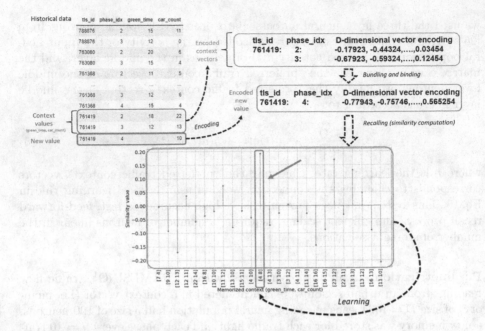

Fig. 3. TRAMESINO system encoding and similarity mechanisms.

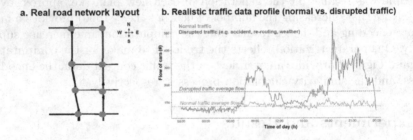

Fig. 4. Real-world road network layout and normal vs. disrupted data.

the velocity which might create jams, whereas, special activities such as football matches or beginning/end of holidays increase the flow magnitude. Using the real-world flow in the dataset and SUMO, we reproduce the traffic flow behavior when disruption occurs starting from normal traffic flow data by reflecting the disruption effect on vehicles speed and/or network capacity and demand. We sweep the disruption magnitude from normal traffic up to 3 levels of disruption (i.e. low, medium, high) reflected over all the 8 crosses over 24 h. The evaluated systems are the following:

- BASELINE: static traffic planning that uses pre-stored timing plans computed offline using historic data.
- MILP: Mixed-Integer Linear Programming traffic optimization implementation inspired from [18].

- HOPFIELD: Hopfield neural network implementation inspired from [16].
- TRAMESINO: instantiation of our system per each traffic light installed in each direction of each of the 8 crosses in the road network.

For the evaluation of the different approaches (i.e. BASELINE, MILP, HOP-FIELD, and TRAMESINO), we followed the next procedure:

- Simulate real-world SPRING-MUSTARD flows in SUMO and store the results (without disruptions and with the 3 levels of progressive disruptions) for each of the five approaches.
- Compute relevant traffic aggregation metrics (i.e. average trip duration, average speed, and waiting time, respectively).
- Rank experiments depending on performance.
- Perform statistical tests (i.e. a combination of omnibus ANOVA and posthoc pairwise T-test with a significance $p = 0.05$) and adjust ranking depending on significance.
- Evaluate best algorithms depending on ranking for subsets of relevant metrics (i.e. the metrics with significant difference).

Our evaluation results are given in Table 1 where each of the approaches is ranked across the disruption magnitude scale (no disruption (N) to max disruption (H)) over the specific metrics (i.e. average trip duration, average speed, and waiting time, respectively). For flow magnitude disruptions, the level of disruption (i.e. low (L), medium (M), and high (H)) is a factor used to adjust the number of vehicles during the disruption. As one can see in Table 1, TRAMESINO overcomes both HOPFIELD and BASELINE, but deviated from the optimal MILP solution with under 30% in average trip duration and waiting time, and just under 3% in average speed. This is due to the optimal solution that MILP finds given the constraints that the values of the traffic quantities rely upon. However, this performance decreases in the typical metrics is successfully compensated by the run-time analysis in Fig. 5. Here, when simulating one day of traffic, TRAMESINO demonstrates that storing and recalling traffic context memories is almost 2× faster than BASELINE and up to 5× faster than MILP, when considering the actual optimization time (i.e. for TRAMESINO store and recall based on similarity, constrained optimization convergence for MILP). Additionally, the overall TRAMESINO processing only took around 12% from the total simulation time of a single day (i.e. approx. 24 min). A specific analysis and evaluation for TRAMESINO is the accuracy and robustness of the high-dimensional encoding. We explored how does the size of the encoding D influence the encoding and decoding of each memory (i.e. in the storing and recall processes of TRAMESINO). Intuitively, a higher dimension of the encoding will support more accurate representations. This is visible in Fig. 6, where we stored and recalled a varying number of traffic data memories of different dimensions. Please recall that this process describes the entire functionality pipeline of TRAMESINO described in Fig. 1.

Table 1. Performance evaluation of the different systems in normal traffic (N) and with varying disruption levels: low (L), medium (M), and high (H). Besides absolute ranking we take the average performance deviation from the optimal solution of MILP.

System/Disruption level	N	L	M	H	Ranking	Deviation
Average trip duration (s)						
BASELINE	168.805	181.217	265.546	270.167	4	49.86%
MILP	118.336	132.406	167.173	167.673	1	0.0%
HOPFIELD	151.281	151.381	223.017	257.464	3	32.28%
TRAMESINO	156.379	157.371	203.775	236.224	2	28.44%
Average speed (km/h)						
BASELINE	58.15	56.78	49.38	47.50	4	10.95%
MILP	59.30	60.00	59.40	59.10	1	0.0%
HOPFIELD	59.48	59.97	49.28	46.18	3	9.84%
TRAMESINO	59.78	59.02	52.08	48.28	2	8.14%
Waiting time (s)						
BASELINE	16.45	18.53	32.59	35.13	4	7.02%
MILP	13.98	16.14	15.14	15.07	1	0.0%
HOPFIELD	13.98	14.96	29.32	37.29	3	5.84%
TRAMESINO	14.95	14.57	22.16	29.01	2	2.96%

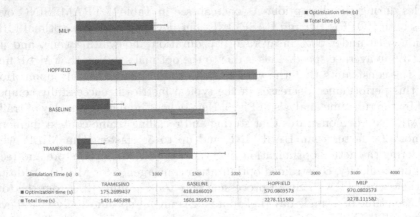

Fig. 5. Run-time performance evaluation for the real-world flows in the simulator.

4   Discussion

Traffic optimization and control is a complex multi-factorial problem. Such a problem requires accurate models, robust control, and, above all, efficient computation, to meet real-world constraints. But, there is a trade-off to be made in order to accommodate all these objectives. Combining human expertise,

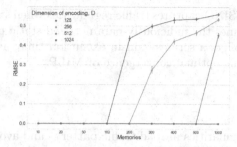

Fig. 6. Encoding/decoding accuracy of TRAMESINO memories.

simple models, and heuristics, the typical static plans (i.e. BASELINE) are the best choice when there is a predictable traffic demand and no dynamic changes in the flow (i.e. accounts for a look-up-table query). Such models fail to capture and accommodate sudden changes in the traffic context and are typically used as fall-back mechanisms. Increasing the price of modelling and computation with mathematical programming and constrained optimization, adaptive systems (i.e. MILP [18]) are the choice for accurate responses to abrupt changes in traffic dynamics. As computing new traffic light plans is required very often (e.g. every 5 min), optimization-based systems reach their limitation at scale, when controlling large urban networks. Constrained optimization might provide the optimal solution but miss the timing. Trying to balance accuracy and computation efficiency, while exploiting the regularity in traffic patterns for robust control, optimization-free methods [16] were developed. Such class of methods, of which TRAMESINO is a member, try to exploit temporal regularities in the traffic data to store relevant patterns of action-consequence (i.e. green time/flow of cars) to be able to bypass expensive optimization.

Looking at the evaluation in Table 1 we see that the accuracy trade-off is visible, optimization-free methods (i.e. HOPFIELD and TRAMESINO) ranking worse than MILP in the traffic specific performance metrics. This is due to the fact that MILP's constrained optimization focuses on satisfying all dependencies among traffic data quantities (i.e. traffic flow, green time, phase offset) in order to provide green time values that minimize trip duration, maximize speed, and reduces waiting time, respectively. The power of such an approach is visible also when progressive disruptions are introduced over the daily traffic patterns. TRAMESINO outperforms the HOPFIELD model due to its efficient computation using NEF and spiking neural networks. This allows for an efficient high-dimensional data representation, simple algebraic operations, and memory dynamics, that can exploit traffic data regularities. These regularities captured by TRAMESINO's memory yield fast adaptation to sudden changes in the traffic flow patterns. As shown in Table 1 despite the increasing disruption levels TRAMESINO is still providing the second best speed, average trip duration, and waiting time. Finally, due to its efficient computation TRAMESINO dominates in terms of run-time (see Fig. 5). Thanks to its learning and adaptation capabilities, TRAMESINO captures traffic regularities when storing new context memories and overcomes the judiciously-parametrized static plan of the BASE-LINE system. This offers a serious gain in execution time, avoiding relaxation of HOPFIELD and the optimal convergence of MILP.

5   Conclusions

In order to exploit regularities in road traffic patterns and avoid expensive optimization techniques, TRAMESINO stands out as a good candidate for efficient traffic control. The system exploits the causal relation among action - consequences pairs (i.e. traffic light green time - flow of cars) in time in order to store relevant contexts. Such traffic context memories are subsequently recalled in new situations, but similar, traffic situations bypassing a new traffic plan re-computation. Our experiments on real-world data demonstrate that such an approach provides a good trade-off between accuracy and robustness overcoming the static plans heuristics and the expensive optimization through a superior gain in run-time. This behavior benefits from the rather deterministic daily traffic profile but, as our experiments demonstrate, can also accommodate sudden disruptions of increasing magnitudes.

References

1. Bracewell, R.N., Bracewell, R.N.: The Fourier Transform and Its Applications, vol. 31999. McGraw-Hill, New York (1986)
2. Day, C.M., Bullock, D.M.: Optimization of traffic signal offsets with high resolution event data. J. Transp. Eng. Part A Syst. **146**(3), 04019076 (2020)
3. Dhamija, S., Gon, A., Varakantham, P., Yeoh, W.: Online traffic signal control through sample-based constrained optimization. In: Proceedings of the International Conference on Automated Planning and Scheduling, vol. 30, pp. 366–374 (2020)
4. Eliasmith, C., Anderson, C.H.: Neural Engineering: Computation, Representation, and Dynamics in Neurobiological Systems. MIT Press, Cambridge (2003)
5. Gayler, R.W.: Vector symbolic architectures answer Jackendoff's challenges for cognitive neuroscience. arXiv preprint cs/0412059 (2004)
6. Henry, J.J., Farges, J.L., Tuffal, J.: The PRODYN real time traffic algorithm. In: Control in Transportation Systems, pp. 305–310. Elsevier (1984)
7. Hoogendoorn, S.P., Bovy, P.H.: Generic gas-kinetic traffic systems modeling with applications to vehicular traffic flow. Transp. Res. Part B Methodol. **35**(4), 317–336 (2001)
8. Hopfield, J.J.: Neurons with graded response have collective computational properties like those of two-state neurons. Proc. Natl. Acad. Sci. **81**(10), 3088–3092 (1984)
9. Hu, H., Liu, H.X.: Arterial offset optimization using archived high-resolution traffic signal data. Transp. Res. Part C Emerg. Technol. **37**, 131–144 (2013)
10. Hunt, P., Robertson, D., Bretherton, R., Royle, M.C.: The scoot on-line traffic signal optimisation technique. Traffic Eng. Contr. **23**(4), 190–192 (1982)

11. Köhler, E., Strehler, M.: Traffic signal optimization: combining static and dynamic models. Transp. Sci. **53**(1), 21–41 (2019)
12. Lendaris, G.G., Mathia, K., Saeks, R.: Linear Hopfield networks and constrained optimization. IEEE Trans. Syst. Man Cybern. Part B (Cybern.) **29**(1), 114–118 (1999)
13. Lopez, P.A., et al.: Microscopic traffic simulation using SUMO. In: The 21st IEEE International Conference on Intelligent Transportation Systems. IEEE (2018). https://elib.dlr.de/124092/
14. Lowrie, P.: Scats, Sydney co-ordinated adaptive traffic system: A traffic responsive method of controlling urban traffic. Roads and Traffic Authority NSW, Traffic Control Section (1990)
15. Mirus, F., Blouw, P., Stewart, T.C., Conradt, J.: An investigation of vehicle behavior prediction using a vector power representation to encode spatial positions of multiple objects and neural networks. Front. Neurorobot. **13**, 84 (2019)
16. Nishikawa, I., Iritani, T., Sakakibara, K.: Improvements of the traffic signal control by complex-valued Hopfield networks. In: The 2006 IEEE International Joint Conference on Neural Network Proceedings, pp. 459–464. IEEE (2006)
17. Nishikawa, I., Kuroe, Y.: Dynamics of complex-valued neural networks and its relation to a phase oscillator system. In: Pal, N.R., Kasabov, N., Mudi, R.K., Pal, S., Parui, S.K. (eds.) ICONIP 2004. LNCS, vol. 3316, pp. 122–129. Springer, Heidelberg (2004). https://doi.org/10.1007/978-3-540-30499-9_18
18. Ouyang, Y., Zhang, R.Y., Lavaei, J., Varaiya, P.: Large-scale traffic signal offset optimization. IEEE Trans. Control Netw. Syst. **7**(3), 1176–1187 (2020)
19. Plate, T.A.: Holographic Reduced Representation: Distributed representation for cognitive structures. CSLI Lecture Notes (2003)
20. Punzo, V., Simonelli, F.: Analysis and comparison of microscopic traffic flow models with real traffic microscopic data. Transp. Res. Rec. **1934**(1), 53–63 (2005)
21. Salort Sánchez, C., Wieder, A., Sottovia, P., Bortoli, S., Baumbach, J., Axenie, C.: GANNSTER: graph-augmented neural network spatio-temporal reasoner for traffic forecasting. In: Lemaire, V., Malinowski, S., Bagnall, A., Guyet, T., Tavenard, R., Ifrim, G. (eds.) AALTD 2020. LNCS (LNAI), vol. 12588, pp. 63–76. Springer, Cham (2020). https://doi.org/10.1007/978-3-030-65742-0_5
22. Sun, J., Liu, H.X.: Stochastic eco-routing in a signalized traffic network. Transp. Res. Procedia **7**, 110–128 (2015)
23. Treiber, M., Kesting, A.: Traffic Flow Dynamics: Data, Models and Simulation. Springer, Heidelberg (2013). https://doi.org/10.1007/978-3-642-32460-4
24. Treiber, M., Kesting, A., Helbing, D.: Understanding widely scattered traffic flows, the capacity drop, and platoons as effects of variance-driven time gaps. Phys. Rev. E **74**(1), 016123 (2006)
25. Voelker, A.R., Crawford, E., Eliasmith, C.: Learning large-scale heteroassociative memories in spiking neurons. Unconv. Comput. Natural Comput. **7**, 2014 (2014)
26. van Wageningen-Kessels, F., van Lint, H., Vuik, K., Hoogendoorn, S.: Genealogy of traffic flow models. EURO J. Transp. Log. **4**(4), 445–473 (2014). https://doi.org/10.1007/s13676-014-0045-5
27. Walsh, M.P., Flynn, M.E., O'Malley, M.J.: Augmented Hopfield network for mixed-integer programming. IEEE Trans. Neural Networks **10**(2), 456–458 (1999)

Detection of Critical Events in Renewable Energy Production Time Series

Laurens P. Stoop[1,2,3]([✉]) (iD), Erik Duijm[1] (iD), Ad Feelders[1] (iD),
and Machteld van den Broek[4] (iD)

[1] Utrecht University, 3584 CS Utrecht, The Netherlands
[2] Royal Netherlands Meteorological Institute, De Bilt, The Netherlands
[3] TenneT TSO B.V., Arnhem, The Netherlands
[4] University of Groningen, 9747 AG Groningen, The Netherlands
l.p.stoop@uu.nl

Abstract. The introduction of more renewable energy sources into the energy system increases the variability and weather dependence of electricity generation. Power system simulations are used to assess the adequacy and reliability of the electricity grid over decades, but often become computational intractable for such long simulation periods with high technical detail. To alleviate this computational burden, we investigate the use of outlier detection algorithms to find periods of extreme renewable energy generation which enables detailed modelling of the performance of power systems under these circumstances. Specifically, we apply the Maximum Divergent Intervals (MDI) algorithm to power generation time series that have been derived from ERA5 historical climate reanalysis covering the period from 1950 through 2019. By applying the MDI algorithm on these time series, we identified intervals of extreme low and high energy production. To determine the outlierness of an interval different divergence measures can be used. Where the cross-entropy measure results in shorter and strongly peaking outliers, the unbiased Kullback-Leibler divergence tends to detect longer and more persistent intervals. These intervals are regarded as potential risks for the electricity grid by domain experts, showcasing the capability of the MDI algorithm to detect critical events in these time series. For the historical period analysed, we found no trend in outlier intensity, or shift and lengthening of the outliers that could be attributed to climate change. By applying MDI on climate model output, power system modellers can investigate the adequacy and possible changes of risk for the current and future electricity grid under a wider range of scenarios.

Keywords: Energy climate · Power system modelling · Outlier detection · Time series · Climate change · Anomaly detection · High impact events

1 Introduction

With the energy transition from fossil-fuel driven generation towards intermittent renewable energy sources like wind and solar power, the electricity

© Springer Nature Switzerland AG 2021
V. Lemaire et al. (Eds.): AALTD 2021, LNAI 13114, pp. 104–119, 2021.
https://doi.org/10.1007/978-3-030-91445-5_7

supply becomes more variable [31]. Additionally, electrification of space heating will enhance [31,39] the already existing variability at the electricity demand side [5,32]. This twofold increase in variability can be partly counteracted by the high interconnectivity of the European electricity system [13] that enables exchanges between countries with either electricity shortfalls or surpluses. However, large scale penetration of variable renewable energy sources can endanger the reliability of the system as weather driven critical conditions may damage elements in the electricity grid or lead to hours with unserved energy [34].

Therefore, insights into critical events are required to support stakeholders with taking appropriate risk reducing investments during the energy transition [40]. For instance, such events could be avoided by investments in flexibility options [14], interconnections [30], storage facilities [24], spatial balancing [16,27] and/or back-up systems.

Power system simulation models can be used to select and quantify these type of investments to deal with critical events in different scenarios of power system development [17]. The simulations often search for cost-effective solutions under pre-set reliability and environmental performance standards. However, when all important features and limitations of the power system are taken into account, these power system simulations can become very complex, resulting in high computational burdens that scale with the simulation period [38].

These constraints on the simulation period impede that power system modellers sufficiently assess the impact of variability of intermittent renewables over different timescales ranging from sub-hourly to decadal [25]. A promising method to comprehensively incorporate the variability of renewables into power system simulations without increasing the simulation period is the Importance Subsampling approach developed by Hilbers et al. [21]. However, this method may overlook important weather-related outliers resulting in an inaccurate assessment of reliability under critical conditions. Although energy system experts could complement the method of Hilbers et al. with information of extreme events in the past [10], such information is lacking for future weather years from climate models [7]. The latter is crucial though, among others for evaluating the power system performance under climate change conditions.

In this paper we apply the Maximally Divergent Intervals (MDI) algorithm developed by Barz et al. [2] that enables the systematic detection of outliers in energy climate datasets, like renewable energy production time series. We perform several experiments on a energy climate dataset of 70 years to determine the merits and limitations of this method to find critical events. The developed method is a key step in a joint project with experts from a national meteorological institute and a Transmission System Operator (TSO). It will be applied to identify critical conditions in very large datasets from climate models to assess system adequacy in many scenarios with power system simulation modelling.

This paper is organized as follows. Related work is discussed in Sect. 2 to place the outlier detection method in a broader context. Section 3 introduces the energy climate dataset used in this study. Next the relevant components of the algorithm are briefly described in Sect. 4. The application of the algorithm is experimentally evaluated and discussed in Sect. 5. Finally, Sect. 6 presents the conclusion and next steps in our project.

2   Related Work

Here we will focus on related work concerned with finding critical events in energy production data and weather data. For related work on algorithms for outlier detection, we refer to the overview in the introduction of Barz et al. [3].

A broad research community addressed the identification of extreme weather events in historical weather years by applying a variation of methods. Where Wu and Chawla [37] focus on using Extreme Value Theory to detect and track heavy rainfall events, others like Duggimpudi et al. [12] used Behavioural outlier Factors to track the path of hurricane Katrina.

Although such extreme weather events may be of interest in their own right due to their potential severity [1], not all high impact events are caused by extreme weather events [33]. Therefore, research is shifting from the identification of extreme weather events to the identification of weather events that have a severe impact on society [40].

The impact based approach asks for a clear definition of a variable that can measure the severity of this impact. Thus searching for weather events that pose a risk for the operation of the power system requires first of all knowledge of how weather influences the power system, secondly a method to classify the weather driven impacts between normal to adverse to severe, and thirdly to detect these severe events. Dawkins and Rushby [10], for example created a composite impact variable capturing the relations between wind droughts and electricity demand peaks. Another example, is the study by van der Wiel et al. [34] who also used a composite variable representing weather dependent solar and wind supply minus the electricity demand. By dividing the renewable generation by the electricity demand, significantly different critical events where found by Drew et al. [11], indicating the importance of the exact definition of the impact variable.

In most of these studies, the impacts are considered severe when the impact value exceeds a pre-defined threshold [10,11,34]. Thus the nature of the impact is pre-determined by this selection criterion and can for example be a shortage or surplus of energy during a specific time horizon. Furthermore, although most studies look at extremes at different time horizons e.g., 1 day, 1 week or 2 weeks, they often fix the length of the time horizon before determining the outliers. As the intensity, duration and/or timing of high impacts can change due to climate change, a more flexible method would be beneficial when looking at climate change related risks.

Finding critical events for the power system thus requires knowledge of the relation between weather and impact. Expert opinion is a way to determine if an event is critical, but it might be very subjective. A thorough overview of critical events for the United Kingdom is given by Dawkins and Rushby [10] where they rely on extensive expert knowledge, and by Ward [35] for the wider region of Europe, though their work could be considered dated given the fast transition.

Additionally, despite the effort of the energy climate community the input data for such studies are not available in a coordinated way for most countries [7]. Using labelled real world data for training an outlier detection method is thus not a viable option, synthetic time series are therefore used within the energy climate community. This limited availability of data is especially an issue with respect to energy consumption data. Methods exist to model the energy consumption [9, 26, 32], but the difficulty in the acquisition of the data required limits the scope of this paper to renewable energy generation.

3 The Energy Climate Dataset

In this section we provide a brief introduction into the data used for our experiments and how it was generated. We first discuss the properties of the ERA5 dataset in Sect. 3.1. After this we will discuss, in Sect. 3.2, the energy conversion models used to create electricity generation data based on the ERA5 reanalysis data.

3.1 The ERA5 Reanalysis Data

ERA5 is the latest reanalysis dataset developed by the European Centre for Medium-Range Weather Forecasts [20]. In a reanalysis dataset [19], historical observations are consistently assimilated into numerical weather models to give a best estimate of the recent climate.

ERA5 reanalysis data stretches from 1950 to the present, with a two month delay. The period between 1950 and 1979 is the preliminary version of the ERA5 back-extension [4]. The ERA5 and its back-extension have undergone significant quality control and are considered state-of-the-art. The variables used in this research are solar irradiance, wind speed at 100 m height, and 2 m temperature.

The temporal granularity of the data is hourly, with a spatial granularity of 0.25° or ±31 km. The period we covered spans from 1950 through 2019. In the spatial domain we used the subregion of Europe, defined here as the region between latitude −14.75 to 40 East and longitude 35 to 74.75 North.

3.2 Energy Conversion Models

To calculate the electricity generation based on climate reanalysis data one needs to know the capacity factor of wind turbines and solar photo-voltaic panels per grid cell, and the distribution of their capacity over the region of interest. The first can be obtained by using conversion models that compute a capacity factor for each grid cell based on the climate variables in that grid cell. The second, a distribution of renewable energy sources for the target year of 2050 was provided to us upon request by Bas van Zuijlen, for details on the properties of this possible distribution we refer to [41].

In collaboration with the TSO stakeholder of our project, several conversion models were compared and analysed. For solar panel electricity generation we compared the methods presented in [23] and [6]. More advanced methods where not used as those require additional information on panel tilt, angle and solar radiation components that are not available. We selected the method as set out by [23], we refer to them for more details.

For wind turbine electricity generation we compared the methods described in [8, 15, 23, 28, 29]. Based on the model complexity, running time and accuracy of the output, we selected the general power curve method from [23]. However, we made three adjustments to this model. First, we reduced the effective capacity factor (CF_e) with 5% to 95% to represent the wake losses in large scale wind-farms. Secondly, we introduce a linear decay in the capacity factor at high wind speeds to more accurately represent high windspeed operational conditions. The third change was that we tuned the power curve regimes. Equation (1) gives the capacity factor for wind turbines (CF_{wind}) used in this study.

$$CF_{wind}(t) = CF_e \times \begin{cases} 0 & \text{if} & V(t) < V_{CI}, \\ \frac{V(t)^3 - V_{CI}^3}{V_R^3 - V_{CI}^3} & \text{if} & V_{CI} \leq V(t) < V_R, \\ 1 & \text{if} & V_R \leq V(t) < V_D, \\ \frac{V_{CO} - V(t)}{V_{CO} - V_D} & \text{if} & V_D \leq V(t) < V_{CO}, \\ 0 & \text{if} & V(t) \geq V_{CO}. \end{cases} \quad (1)$$

Here $V(t)$ is the wind speed at the height of the wind turbine and the power curve regimes are given by the cut-in ($V_{CI} = 3\,\text{m/s}$), rated ($V_R = 11\,\text{m/s}$), decay ($V_D = 20\,\text{m/s}$) and cut-out ($V_{CO} = 25\,\text{m/s}$) wind speed. The windspeed provided by ERA5 (at 100 m) did not match the hub height for offshore turbines within the capacity distribution used [41], therefore it is scaled using the wind profile power law to 150 m. The surface roughness was set to a constant value for both onshore ($\alpha = 0.143$) and offshore regions ($\alpha = 0.11$).

The total energy generation per grid cell is obtained by multiplying the capacity factor with the installed capacity from the distribution used.

The temporal variations in supply are expected to play a larger role than the spatial variation for the critical conditions in the power system. Additionally, the current European electricity grid is highly interconnected[1], even higher interconnectivity of the system is expected by 2050. As we search for critical conditions and we have to reduce the dataset size for tractability, we assume that the electricity grid can be approximated by a copper plate [41]. This implies that the flow of electricity is not impeded and local inbalances are dealt with on system wide scale.

Due to the copperplate assumption we can sum the electricity generation per technology over the European region to obtain time series data. Our final input time series data thus contains three variables, namely wind-onshore (WON), wind-offshore (WOF), and solar photo-voltaic (SPV) electricity generation. This data is based on historical weather years (1950–2019), but uses a possible distribution of renewables in a deep decarbonised future. For each variable the length of the time series is therefore $N = 613,594$.

---

[1] See https://www.entsoe.eu/data/map/ for a interactive map of the current network.

4 The MDI Algorithm

In this section we give a short description of the Maximally Divergent Intervals (MDI) algorithm (see [2,3] for more details). This algorithm finds outliers in spatial-temporal data, but since we aggregate over the spatial component, we will restrict the presentation to the strictly temporal case. Let

$$\{(x_{t,1}, x_{t,2}, \ldots, x_{t,d}) : t = 1, \ldots, N\}$$

be a d-dimensional multivariate time series of length N. Individual samples are written as $\mathbf{x}_t \in \mathbb{R}^d$. Loosely speaking, an outlier is an interval in which the distribution of the variables deviates strongly from their distribution outside that interval. To model the probability distribution, Kernel Density Estimation (KDE) using Gaussian kernels or a multivariate Gaussian distribution are applied. The anomaly score of interval I is defined as:

$$S(I) = \mathcal{D}(\hat{p}_I, \hat{p}_\Omega), \qquad I \in \mathcal{I} \tag{2}$$

where $\mathcal{D}$ is some measure of the divergence between two probability distributions, $\hat{p}_I$ is the distribution fitted to the observations inside the interval, and $\hat{p}_\Omega$ is the distribution fitted to the remaining observations. The set $\mathcal{I}$ contains all intervals with a time horizon length between a user-specified minimum a and maximum b, hence $|\mathcal{I}| \approx N(b - a)$. Possible divergence measures are cross-entropy, (unbiased) Kullback-Leibler and Jensen-Shannon divergence. Although Jensen-Shannon divergence has its merits, it was found not to be tractable due to the size of our data. The cross-entropy and Kullback-Leibler divergence are respectively computed by:

$$\mathcal{D}_{\text{CE}}(I, \Omega) = \frac{1}{|I|} \sum_{t \in I} \log \hat{p}_\Omega(\mathbf{x}_t), \text{ and } \mathcal{D}_{\text{KL}}(I, \Omega) = \frac{1}{|I|} \sum_{t \in I} \log \left(\frac{\hat{p}_I(\mathbf{x}_t)}{\hat{p}_\Omega(\mathbf{x}_t)} \right),$$

where $\hat{p}_I(\mathbf{x}_t)$ is the probability density of data point t according to the probability density fitted to the interval, and likewise $\hat{p}_\Omega(\mathbf{x}_t)$ is the probability density of data point t according to the probability density fitted to the remainder of the data. Barz et al. [3] note that $\mathcal{D}_{\text{KL}}$ has a bias towards smaller intervals, and propose the unbiased variant $\mathcal{D}_{\text{U-KL}} = 2 \cdot |I| \cdot \mathcal{D}_{\text{KL}}$. If a multivariate Gaussian distribution is used to estimate the probability densities, then the (unbiased) Kullback-Leibler divergence can be computed quite efficiently, since in that case a closed-form solution is available.

To take into account the temporal correlation between data points, a technique call time-delay embedding is applied. Time-delay embedding incorporates context from previous time-steps into each sample by transforming a given time series $\{\mathbf{x}_t\}_{t=1}^N, \mathbf{x}_t \in \mathbb{R}^d$ into another time-series $\{\mathbf{x}_t'\}_{t=1+(\kappa-1)\tau}^N, \mathbf{x}_t' \in \mathbb{R}^{\kappa d}$, with

$$\mathbf{x}_t' = \left(\mathbf{x}_t^\top \ \mathbf{x}_{t-\tau}^\top \ \mathbf{x}_{t-2\tau}^\top \ \cdots \ \mathbf{x}_{t-(\kappa-1)\tau}^\top \right)^\top .$$

Here the embedding dimension κ specifies the number of samples to stack together and the time lag τ specifies the gap between two consecutive time-steps to be included as context.

To make the algorithm better suited for large data sets, a method that proposes intervals that are likely to contain outliers is used. The idea behind the method is that an outlier interval tends to contain several data points that would receive high scores when using point wise outlier detection. One such point wise scoring method is Hotelling's T^2 score [22] (or squared Mahalanobis distance):

$$T_t^2 = (\mathbf{x}_t - \hat{\boldsymbol{\mu}})^\top \hat{\boldsymbol{\Sigma}}^{-1} (\mathbf{x}_t - \hat{\boldsymbol{\mu}}).$$

At the start and end of an outlying interval, respectively, an increase and decrease of the point wise scores is expected. Therefore, only intervals that start and end with data points whose

$$g(t) = |T_{t+1}^2 - T_{t-1}^2|$$

value surpass the threshold $\theta_g = \hat{\mu}_g + \vartheta \cdot \hat{\sigma}_g$ are considered, where ϑ is a parameter to be set by the user. Thus much less intervals need to be checked leading to a substantial speed up, since estimating distributions and divergence calculations are very time consuming. The potential downside of this approach is that outlier intervals may be overlooked, thus lowering recall. However, experiments performed by Barz et al. [3] show that this was not the case when a reasonable value for ϑ was selected.

In order to ensure that the top detected outliers aren't all small variations of the same event, starting with the top outlier, the overlap:

$$O(I_1, I_2) = \frac{|I_1 \cap I_2|}{|I_1 \cup I_2|}$$

with lower scoring outliers is checked. If this overlap is larger than a user-defined threshold θ_o, only the interval with the higher score is reported. Finally, the algorithm sorts the intervals in descending order of their score, so that a user-specified number of top k intervals can be selected as output.

5   Experimental Results

To determine whether the MDI algorithm is suited to identify critical events in energy climate data we performed several experiments. Each experiment represents a potential use case for our project and partners, while they are also a test case for the tuning and pre-processing used. The outliers found where presented to subject matter experts to determine if they provide insight in critical events that could influence the future energy system.

All experiments are performed on an Intel Xeon Gold 6130 CPU with 16 dual-cores at 2.1 GHz clock speed. Our setup has 125.6 GB of available RAM memory. The multi-threading was limited to using 30 threads.

The rest of this section is organised as follows. First we discuss the tuning performed to make the MDI algorithm usable for renewable energy generation time series data in Sect. 5.1. The top 20 outliers detected using Cross Entropy and the unbiased Kullback-Leibler divergence are then investigated in Sect. 5.2. Finally in Sect. 5.3, we investigate if there are climate change induced changes in the intensity, time of the year and length of the top 50 outliers per decade.

5.1   Tuning of the MDI Algorithm

The settings of the algorithm were chosen in consultation with the domain experts, the model choices presented below are the end result.

Because the single Gaussian distribution gave quite a bad fit, we selected KDE using Gaussian kernels (with kernel width 1) to estimate the probability distributions. Hotelling's T^2 proposal method is used with $\vartheta = 1.5$. The allowed overlap between intervals was set to $\theta_o = 0.5$. The built-in data normalization method of the MDI algorithm, subtraction of the mean and division by the maximum, was used. We used both Cross Entropy and the unbiased Kullback-Leibler divergence to score intervals.

The interval length was set to 2 days minimum, and 10 days maximum. The reason was two-fold, the usefulness of the output as deemed by our experts and tractability of the algorithm. At shorter timescales batteries and the shifting of demand can be utilised to mitigate the effect of an outlier. At longer timescales (sub-)seasonal storage, like hydrodams and hydrogen, can be utilised. However, for the period between 2 and 10 days there are multiple technologies that could be utilised, some of which are not yet fully developed. Knowledge of outliers within this window can therefore help determine what technologies could be utilised or should be further developed. Using a minimum interval length also makes sure there is sufficient data to reliably estimate a distribution.

In order to accurately discover temporal outliers, the temporal context embedding parameters need to be investigated. The idea behind the temporal context embedding is to pick points that are correlated at different time-lags. To investigate the autocorrelation length, the partial autocorrelation per variable was calculated (see Fig. 1). Based on these calculations we have decided to use $\kappa = 4$ and $\tau = 8$ as respectively temporal embedding dimension and time lag settings, as these capture most of the autocorrelation in all variables. They ensure that the autocorrelation in solar photo-voltaic power and onshore wind power at the larger lags of approximately 24 h are accounted for. These settings also ensure that at least one day and night cycle is embedded as context, which has a big impact on the Solar Photovoltaic energy generation in particular.

The original MDI algorithm of Barz et al. [3] is implemented in an open source library[2] with both a `Python` implementation of the algorithm and a `C++` implementation. As the `Python` algorithm is not well suited for large data sets, we used the `C++` implementation and additionally built a wrapper in `Python` that accessed the `C++` multi-threading functionality and added `xarray` compatibility.

---

[2] https://github.com/cvjena/libmaxdiv.

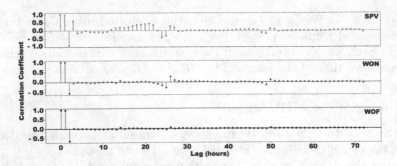

Fig. 1. The autocorrelation of variables at different time lags, using the Yule-Walker method with sample size adjustment.

5.2   Outlier Identification and Assessment

In order to determine if MDI can find potential shortfalls or surges that might affect the European energy system, we investigated the outliers that were identified by two divergence measures. The top outliers detected using Cross Entropy and the unbiased Kullback-Leibler divergence are shown in Figs. 2 and 3, respectively. The top 20 outliers were also presented to our domain experts to harness their insight in the tuning and assessment process. Both the Cross Entropy and unbiased Kullback-Leibler methods took just over 29 h wall clock time to calculate.

According to the domain experts, the top 20 outliers found are all likely to be high impact events. Additional investigation revealed that the top outlier based on Cross Entropy coincides with a period that was identified by Dawkins et al. [10] as an adverse weather system for the electricity system of the United Kingdom and Europe. For the top outlier detected using the unbiased Kullback-Leibler divergence, a historical high impact event was found in the Burns' day storm (25th jan 1991). This storm is considered one of the worst storms of the last century for the United Kingdom, the Netherlands, and Belgium in which 97 people lost their lives.

To summarize them, the top 20 outliers were grouped based on the month in which they occur, the length of the outlier and their type. The type of an outlier is based on three indicators, namely Peak, Trough and Peak-Trough (see Table 1). During a Peak, the power generation is above normal for two or more of the three energy sources. In a Trough, power generation is below normal. The Peak-Trough type indicates that the outlier contains a variable that has a peak as well as one that has a trough, and the combined energy generation over the period is neither very high nor very low.

Based on the grouping we defined classes for the outliers. For the unbiased Kullback-Leibler divergence these classes are Winter Surplus and Summer Deficiency. We consider the outliers that show a peak in the extended winter from November through March to be part of the Winter Surplus class. Outlier events with a trough in overall electricity generation in the extended summer period, from May through September, are part of the Summer Deficiency class. For Cross Entropy we have similar classes: Winter Surplus, Long Term Summer Deficiency and Short Term Summer Deficiency. The only distinction is that for the Summer Deficiency we have sub classes based on the length of the event: outliers that last between 48 and 72 h are considered short term, and outlier events longer than 72 h are considered long term.

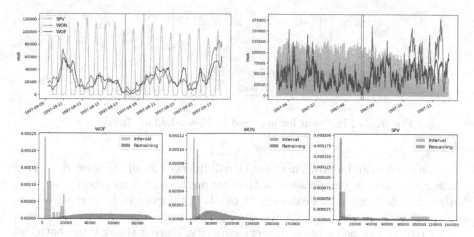

Fig. 2. Figures depicting the outlier with the highest score using the Cross Entropy measure. The top figures show the generation of each technology and the temporal context in which the outlier (indicated by red lines) was found. The bottom images provide histograms of the generation (in MWh) of each of the three technologies during the interval (in their respective colour) and the remaining data (in purple). (Color figure online)

Table 1. Grouping of the top 20 outliers found by the MDI algorithm in our time series data using Cross Entropy and unbiased Kullback-Leibler method. The grouping has been ordered in such a way that the different outlier classes can be discerned easily. It should be noted that although the length of the intervals is near the bounds, they are not at the bounds in general.

Cross Entropy							unbiased Kullback-Leibler						
Top-k	Month	Length(h)	SPV	WON	WOF	Type	Top-k	Month	Length(h)	SPV	WON	WOF	Type
1/6/13/19	Aug.	48-72	+	−	−	T	1/5/7/10	Jan.	216+	−	+	+	P
3/5	June	48-72	+	−	−	T	2/8/17	Dec.	216+	−	+	+	P
07/09/2017	July	48-72	+	−	−	T	3/4	Feb.	216+	−	+	+	P
16	July	72-96	+	−	−	T	11/18-20	Nov.	216+	−	+	+	P
10/15	July	150-175	+	−	−	T	9	Jan.	192 − 216	−	+	+	P
14	Feb.	48-72	−	+	−	PT	6/13/16	Feb.	216+	0	+	+	P
4	Apr.	48-72	0	+	+	P	14	Aug.	216+	+	−	−	T
2/11	Dec.	48-72	−	+	+	P	15	July	216+	+	−	−	T
12/18	Feb.	48-72	−	+	+	P							
20	Jan.	48-72	−	+	+	P							

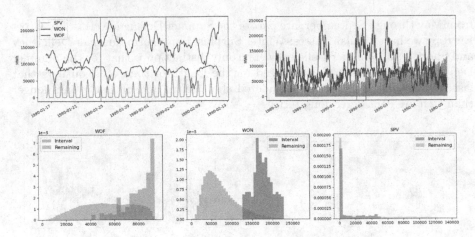

Fig. 3. As Fig. 2, but for unbiased Kullback-Leibler divergence.

These classes can be problematic as they influence the whole network. A long deficiency needs to be compensated with other methods of non-carbon generation that need to be flexible and can be controlled, as the current battery capacities aren't sufficient. Shorter deficiencies during the summer are also problematic, as they require extensive use of battery capacity. During the day the batteries charge on the available solar photo-voltaic energy generation, but at night they need to be discharged to compensate for the lack of wind. This strain on the batteries causes them to wear. An increase in such short deficiencies represents an economic risk, as the batteries would need to be replaced more frequently. The Winter Surplus increases the energy generation of the grid, causing a surplus, which can be problematic if this isn't controlled. The surplus needs to be discharged somehow. This discharge of unused energy represents an economic risk, as the wind turbines and solar panels are wearing down, without the energy that is generated being used.

Based on the top 20 outliers we note that the outliers detected by the Cross Entropy measure tend to have a very short duration, whereas the outliers detected by the unbiased Kullback-Leibler divergence tend to be longer. As a quick reminder, Cross Entropy is related to the Kullback-Leibler divergence measure and the latter was found by Barz et al. [3] to have a bias towards smaller intervals. We can thus expect this tendency to shorter intervals for Cross Entropy outliers. However, the tendency towards longer intervals is unexpected for the unbiased Kullback-Leibler divergence measure as it was created specifically to be unbiased towards interval length. It should be noted that while some outliers are found on the bounds set on the outlier duration, they are in general not on these bounds.

Irrespective of the tendency to be near the boundary interval lengths, both divergence measure studies where deemed to identify likely high impact events by our domain experts. Therefore both measures should be considered when studying high impact events in energy climate data.

5.3   Historic Climate Change and Decadal Variability

The change of intensity, time of the year, and length of outliers might change the impact of an event and is therefore important to consider [17,33]. For these experiments we combined offshore wind, onshore wind and solar photo-voltaic power generation into a single variable called Total Electricity Generation (TEG). This single aggregate provides a reasonable indication of shortages and surges in the electricity system, while reducing the computational burden of the algorithm. We identify the top 50 outliers per decade and use these in our assessment of the intensity, time and length of the outliers over the historic period.

We found that the outliers in the TEG time series represent mostly peaks. Trough-type outliers were difficult to detect in the TEG dataset, especially when using the Cross Entropy measure. Potentially risky situations as in Fig. 2 remain undetected in this univariate analysis. This underlines the added insight provided by the multivariate analysis, and highlights the importance of selecting the correct divergence measure.

The intensity of the outliers is investigated by looking at the average energy generation during the outlier. Figure 4 shows a boxplot of the average Total Energy Generation during the outlier for the top 50 outliers found with Cross Entropy divergence. While there is no linear trend visible, some periodical behaviour appears to influence the outlier events. This periodic behaviour appears in all combinations of top number of outliers investigated and divergence measures used. Due to the presence of Trough-type outlying events this effect is hard see for the unbiased Kullback-Leibler divergence (figure not shown). Similar behaviour of multidecadel variability in German wind energy generation was found by Wohland et al. [36].

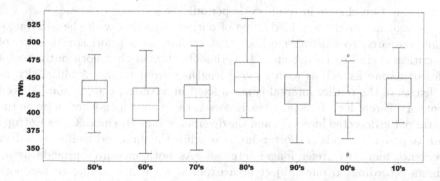

Fig. 4. Boxplot of the average hourly Total Energy Generation during the top 50 outlier events per decade based on the Cross Entropy measure.

These result emphasise that the multidecadel variability needs to be taken into account by policy makers as it influences the strength of the outliers. Assessments of the energy system currently only use a limited set of weather years and might therefore under- or overestimate the extremeness of critical conditions for the energy system.

We studied the timing and duration of the outliers found per decade in the TEG time series to determine whether they are affected by climate change. We did, however, not find any obvious trends or shifts that could potentially be attributed to climate change. Such trends or shifts are possibly masked by the multidecadel variability in the outliers. The time of emergence of a climate change signal lies thus in the future, like it currently does for most climate related impacts [18].

6   Conclusion and Future Work

Using the Maximally Divergent Intervals (MDI) algorithm we found outlying time periods in 70 years of historic weather-derived energy production data on three types of renewable energy. According to subject area experts from a national Transmission System Operator (TSO), the identified outliers indeed represented periods during which the European electricity system could be at risk. However, when the three renewable energy generation variables were combined into a single variable, Total Energy Generation, potential outliers were missed as mostly peak-type outliers were detected. The multivariate analysis is therefor preferred in further work. We conclude that, with the proper parameter settings, outlier detection with MDI can help the assessment of the future energy grid by highlighting the most extreme situations.

When analysing the Total Energy Generation peer decade we found that the intensity of outliers manifests multidecadel variability over the last 70 years. However, we found no trend could be attributed to climate change. This variability in the outliers also hinders the determination of climate change attributable shift or duration change in the historic period.

We demonstrated the added value of outlier detection with the MDI algorithm compared to existing methods that require an a priori specification of the critical events to be detected. Experiments showed that both outliers of a different nature as well as with varying lengths were detected. Additionally, as the length of the outlier interval is not a fixed in advance, comparison between events of different lengths is possible. However, there is a dependency between the lengths of the detected intervals, and the divergence measure used. Cross Entropy tends to prefer intervals of shorter duration, while the unbiased Kullback-Leibler divergence tends to prefer longer intervals. As both measures provide useful insights according to our subject area experts, we will continue to use both measures for outlier detection in energy climate data.

In the next phase of the project, the method presented here will be used for two applications related to the assessment of power system adequacy. First, when the outliers identified are combined with a method to represent the generic variability of the weather, a synthetic representative time series could be constructed. Power system simulations based on a synthetic time series can be used to ensure both representativeness with respect to critical climate conditions as well as computational tractability. Second, besides applying the MDI method on a historical climate dataset as was demonstrated in this paper, it can be applied to climate projections from a multitude of climate simulation models to investigate how climate change and multidecadal climate variability influence the character and frequency of critical conditions for the electricity grid.

For both applications, the method is preferably applied to a dataset that also takes electricity demand into account. For this purpose, the temperature dependant component of the electricity demand should be based on climate variables used for the calculation of the electricity generation from renewable sources. Incorporation of energy consumption data might decrease or exacerbate the impact of critical weather events.

Acknowledgements. The data used in the experiments contains modified Copernicus Climate Change Service information, doi.org/10.24381/cds.adbb2d47 (2020). This research received funding from the Netherlands Organisation for Scientific Research (NWO) under grant number 647.003.005. The methodology presented here was developed as part of the IS-ENES3 project that has received funding from the European Union's Horizon 2020 research and innovation programme under grant agreement No. 824084.

References

1. Arent, D.J., et al.: Key economic sectors and services. In: Climate Change 2014 Impacts, Adaptation and Vulnerability: Part A, pp. 659–708 (2015)
2. Barz, B., Garcia, Y.G., Rodner, E., Denzler, J.: Maximally divergent intervals for extreme weather event detection. In: OCEANS 2017-Aberdeen. IEEE (2017). https://doi.org/10.1109/OCEANSE.2017.8084569
3. Barz, B., Rodner, E., Garcia, Y.G., Denzler, J.: Detecting regions of maximal divergence for spatio-temporal anomaly detection. IEEE Trans. Pattern Anal. Mach. Intell. **41**(5), 1088–1101 (2018). https://doi.org/10.1109/TPAMI.2018.2823766
4. Bell, B., et al.: ERA5 monthly averaged data on single levels from 1950 to 1978. Climate Data Store (CDS) (2020). https://cds.climate.copernicus-climate. eu/. Accessed 10 Nov 2020
5. Bessec, M., Fouquau, J.: The non-linear link between electricity consumption and temperature in Europe: a threshold panel approach. Energy Econ. **30**(5), 2705–2721 (2008). https://doi.org/10.1016/j.eneco.2008.02.003
6. Bett, P.E., Thornton, H.E.: The climatological relationships between wind and solar energy supply in Britain. Renew. Energy **87**, 96–110 (2016). https://doi.org/ 10.1016/j.renene.2015.10.006
7. Bloomfield, H.C., et al.: The importance of weather and climate to energy systems: a workshop on next generation challenges in energy-climate modeling. Bull. Am. Meteorol. Soc. **102**(1), E159–E167 (2021). https://doi.org/10.1175/BAMS-D-20-0256.1
8. Carrillo, C., Obando Montaño, A.F., Cidrás, J., Díaz-Dorado, E.: Review of power curve modelling for windturbines. Renew. Sustain. Energy Rev. **21**, 572–581 (2013). https://doi.org/10.1016/j.rser.2013.01.012

9. Cassarino, T.G., Sharp, E., Barrett, M.: The impact of social and weather drivers on the historical electricity demand in Europe. Appl. Energy **229**, 176–185 (2018). https://doi.org/10.1016/j.apenergy.2018.07.108

10. Dawkins, L., Rushby, I.: Characterising adverse weather for the UK electricity system (2021). nic.org.uk/app/uploads/MetOffice-Characterising-Adverse-Weather-Phase-2a.pdf

11. Drew, D.R., et al.: Sunny windy sundays. Renew. Energy **178**, 870–875 (2019). https://doi.org/10.1016/j.renene.2019.02.029

12. Duggimpudi, M.B., Abbady, S., Chen, J., Raghavan, V.V.: Spatio-temporal outlier detection algorithms based on computing behavioral outlierness factor. Data Knowl. Eng. **122**, 1–24 (2019). https://doi.org/10.1016/j.datak.2017.12.001

13. ENTSO-E: Ten-year network development plan 2020. Technical Report, European Network of Transmission System Operators for Electricity, Brussels (2021)

14. Frew, B.A., Becker, S., Dvorak, M.J., Andresen, G.B., Jacobson, M.Z.: Flexibility mechanisms and pathways to a highly renewable us electricity future. Energy **101**, 65–78 (2016). https://doi.org/10.1016/j.energy.2016.01.079

15. Gonzalez, A., et al.: EMHIRES dataset Part I: Wind power generation (2016). https://doi.org/10.2790/831549

16. Grams, C.M., Beerli, R., Pfenninger, S., Staffell, I., Wernli, H.: Balancing Europe's wind-power output through spatial deployment informed by weather regimes. Nat. Climate Change **7**(8), 557–562 (2017). https://doi.org/10.1038/nclimate3338

17. Harang, I., Heymann, F., Stoop, L.P.: Incorporating climate change effects into the European power system adequacy assessment using a post-processing method. Sustain. Energy Grids Networks **24**, 100403 (2020). https://doi.org/10.1016/j.segan.2020.100403

18. Hawkins, E., Sutton, R.: Time of emergence of climate signals. Geophys. Res. Lett. **39**(1) (2012). https://doi.org/10.1029/2011GL050087

19. Hersbach, H., et al.: Climate data store: ERA5 hourly data on single levels (2018). https://doi.org/10.24381/cds.adbb2d47

20. Hersbach, H., et al.: The ERA5 global reanalysis. Q. J. R. Meteorol. Soc. **146**(730), 1999–2049 (2020). https://doi.org/10.1002/qj.3803

21. Hilbers, A.P., Brayshaw, D.J., Gandy, A.: Importance subsampling: improving power system planning under climate-based uncertainty. Appl. Energy **251**, 113114 (2019). https://doi.org/10.1016/j.apenergy.2019.04.110

22. Hotelling, H.: The generalization of student's ratio. In: Kotz, S., Johnson, N.L. (eds.) Breakthroughs in Statistics. SSS (Perspectives in Statistics). Springer, New York (1992). https://doi.org/10.1007/978-1-4612-0919-5_4

23. Jerez, S., et al.: The CLIMIX model: a tool to create and evaluate spatially-resolved scenarios of photovoltaic and wind power development. Renew. Sustain. Energy Rev. **42**, 1–15 (2015). https://doi.org/10.1016/j.rser.2014.09.041

24. Kies, A., Schyska, B.U., von Bremen, L.: The effect of hydro power on the optimal distribution of wind and solar generation facilities in a simplified highly renewable European power system. Energy Procedia **97**, 149–155 (2016). https://doi.org/10.1016/j.egypro.2016.10.043

25. McCollum, D.L., Gambhir, A., Rogelj, J., Wilson, C.: Energy modellers should explore extremes more systematically in scenarios. Nat. Energy **5**(2), 104–107 (2020). https://doi.org/10.1038/s41560-020-0555-3

26. Moral-Carcedo, J., Vicéns-Otero, J.: Modelling the non-linear response of Spanish electricity demand to temperature variations. Energy Econ. **27**(3), 477–494 (2005). https://doi.org/10.1016/j.eneco.2005.01.003

27. Neubacher, C., Witthaut, D., Wohland, J.: Multi-decadal offshore wind power variability can be mitigated through optimized European allocation. Adv. Geosci. **54**, 205–215 (2021). https://doi.org/10.5194/adgeo-54-205-2021
28. Ruiz, P., et al.: ENSPRESO - an open, EU-28 wide, transparent and coherent database of wind, solar and biomass energy potentials. Energy Strategy Rev. **26**, 100379 (2019). https://doi.org/10.1016/j.esr.2019.100379
29. Saint-Drenan, Y.M., et al.: A parametric model for wind turbine power curves incorporating environmental conditions. Renew. Energy **157**, 754–768 (2020). https://doi.org/10.1016/j.renene.2020.04.123
30. Schlachtberger, D., Brown, T., Schramm, S., Greiner, M.: The benefits of cooperation in a highly renewable European electricity network. Energy **134**, 469–481 (2017). https://doi.org/10.1016/j.energy.2017.06.004
31. Staffell, I., Pfenninger, S.: The increasing impact of weather on electricity supply and demand. Energy **145**, 65–78 (2018). https://doi.org/10.1016/j.energy.2017.12.051
32. Thornton, H., Hoskins, B.J., Scaife, A.: The role of temperature in the variability and extremes of electricity and gas demand in great Britain. Environ. Res. Lett. **11**, 114015 (2016). https://doi.org/10.1088/1748-9326/11/11/114015
33. van der Wiel, K., Selten, F.M., Bintanja, R., Blackport, R., Screen, J.A.: Ensemble climate-impact modelling: extreme impacts from moderate meteorological conditions. Environ. Res. Lett. **15**, 034050 (2020). https://doi.org/10.1088/1748-9326/ab7668
34. van der Wiel, K., Stoop, L.P., van Zuijlen, B.R.H., Blackport, R., van den Broek, M.A., Selten, F.M.: Meteorological conditions leading to extreme low variable renewable energy production and extreme high energy shortfall. Renew. Sustain. Energy Rev. **111**, 261–275 (2019). https://doi.org/10.1016/j.rser.2019.04.065
35. Ward, D.M.: The effect of weather on grid systems and the reliability of electricity supply. Climatic Change **121**(1), 103–113 (2013). https://doi.org/10.1007/s10584-013-0916-z
36. Wohland, J., Omrani, N.E., Keenlyside, N., Witthaut, D.: Significant multidecadal variability in German wind energy generation. Wind Energy Sci. **4**(3), 515–526 (2019). https://doi.org/10.5194/wes-4-515-2019
37. Wu, E., Chawla, S.: Spatio-temporal analysis of the relationship between south American precipitation extremes and the el niño southern oscillation. In: ICDMW 2007 (2007). https://doi.org/10.1109/ICDMW.2007.102
38. Wuijts, R., van den Broek, M., van den Akker, J.: Effect of modeling choices in the unit commitment problem. Applied Energy (2021, Submitted)
39. Zeyringer, M., Price, J., Fais, B., Li, P.H., Sharp, E.: Designing low-carbon power systems for great Britain in 2050 that are robust to the spatiotemporal and interannual variability of weather. Nat. Energy **3**(5), 395–409 (2018). https://doi.org/10.1038/s41560-018-0128-x
40. Zscheischler, J., van den Hurk, B., Ward, P.J., Westra, S.: Multivariate extremes and compound events. In: Climate Extremes and their Implications for Impact and Risk Assessment. Elsevier (2020)
41. van Zuijlen, B., Zappa, W., Turkenburg, W., van der Schrier, G., van den Broek, M.: Cost-optimal reliable power generation in a deep decarbonisation future. Appl. Energy **253**, 113587 (2019). https://doi.org/10.1016/j.apenergy.2019.113587

Poster Presentation

Multimodal Meta-Learning for Time Series Regression

Sebastian Pineda Arango[1(✉)], Felix Heinrich[2], Kiran Madhusudhanan[1],
and Lars Schmidt-Thieme[1]

[1] University of Hildesheim, Hildesheim, Germany
pineda@uni-hildesheim.de,
{kiranmadhusud,schmidt-thieme}@ismll.uni-hildesheim.de
[2] Volkswagen AG, Wolfsburg, Germany
felix.heinrich1@volkswagen.de

Abstract. Recent work has shown the efficiency of deep learning models such as Fully Convolutional Networks (FCN) or Recurrent Neural Networks (RNN) to deal with Time Series Regression (TSR) problems. These models sometimes need a lot of data to be able to generalize, yet the time series are sometimes not long enough to be able to learn patterns. Therefore, it is important to make use of information across time series to improve learning. In this paper, we will explore the idea of using meta-learning for quickly adapting model parameters to new short-history time series by modifying the original idea of Model Agnostic Meta-Learning (MAML) [3]. Moreover, based on prior work on multimodal MAML [22], we propose a method for conditioning parameters of the model through an auxiliary network that encodes global information of the time series to extract meta-features. Finally, we apply the data to time series of different domains, such as pollution measurements, heart-rate sensors, and electrical battery data. We show empirically that our proposed meta-learning method learns TSR with few data fast and outperforms the baselines in 9 of 12 experiments.

Keywords: Meta-learning · Time series regression · Meta-features extraction

1 Introduction

Time series regression is a common problem that appears when hidden variables should be inferred given a known multivariate time series. It finds applicability on a broad range of areas such as predicting heart-rate, pollution levels or state-of-charge of batteries. However, in order to train a model with high accuracy,

---

S. P. Arango and F. Heinrich—Equal contribution.

Electronic supplementary material The online version of this chapter (https://doi.org/10.1007/978-3-030-91445-5_8) contains supplementary material, which is available to authorized users.

V. Lemaire et al. (Eds.): AALTD 2021, LNAI 13114, pp. 123–138, 2021.
https://doi.org/10.1007/978-3-030-91445-5_8

a lot of data is needed. This sets some practical limitations, for instance, when a model is to be deployed on a new system with unknown conditions such as a new user or an older state of a battery, as new conditions cause a domain shift, to whom a model should be adapted based on few historical data.

On the other hand, recent work on meta-learning for image classification has shown that it is possible to achieve fast adaptation when having few data [3]. The core idea is to learn how to adapt the parameters efficiently by looking at many mutually-exclusive and diverse classification tasks [16]. Nevertheless, deriving a lot of tasks requires special task designs. In image classification, sampling classes and shuffling the labels have enabled these diverse tasks.

In time series regression problems, it is usual to have very long but few time series, as every time series is generated from a specific and small set of conditions (e.g. different subjects, machines, or cities). Therefore, applying powerful ideas from meta-learning is not trivial. A proof of it is the fact that there have been few works on how to apply those methods to time series regression, or even related problems such as time series forecasting.

The central idea of this work is to extend model-agnostic meta-learning (MAML) [3] and multi-modal MAML (MMAML) [22] to time series regression. For that, we define the meta-learning problem as how to adapt fast to a new task given a set of known samples, namely *support set*, so that we perform well on predicting the output channel for other samples that belong to the same task, or *query set*.

We summarize the contributions from our paper as:

- We propose a specific method for generating diverse tasks by assuming the real scenario of few but long time series at hand.
- It is the first work on how to extend MAML and MMAML to time series regression, and that demonstrates its improved performance over transfer learning.
- We show the utility of our ideas through empirical evaluations on three datasets and compare with different baselines by proposing an evaluation protocol that can be applied for future work on meta-learning for TSR.

2   Related Work

In recent years, there has been a lot of work on meta-learning applied on few-shot settings, specially in problems related to image classification and reinforcement learning [3,13,19,22]. All of them share some commonalities, such as, an inner loop, or so-called *base learner* that aims to use the support set to adapt the model parameters, and an outer loop, or *meta-learner*, that modifies the base-learner meta-parameters so that it learns faster. Sometimes the meta-learner is another model such as LSTM [17] or includes memory modules [18]. Other approaches learn a metric function that allows finding fast a good prototype given a few samples of a new task [19].

Nevertheless, among the landscape of all the methods for performing meta-learning, MAML stands out because it does not include additional parameters, but just aims to learn a good initialization so that a model achieves good performance after few gradient descent updates. Some methods, such as TADAM [13] or Multimodal MAML (MMAML) [22] extend MAML by using additional networks for embedding the whole task and conditioning the parameters of the predictive model. Other methods simplify the optimization introduced in MAML by using first-order approximations to avoid computing Hessians [10].

It is possible to find previous work on meta-learning time series for specific problems. Lemke et al. [8] propose a model that can learn rules on how to apply models on the Time Series Forecasting (TSF) problem. For that, they extract different features from the time series such as Kurtosis and Lyapunov coefficient. Similarly, Talagala et al. [20] train a Random Forest model that decides which is the best model to use on a new TSF problem. However, selecting the best model or creating rules for it does not yield a continuous search space. Given a specific support set, the set of possible models is discrete, and therefore, very limited. N-BEATS [12] is presented as a meta-learning option for zero-shot learning TSF that achieves good performance under unseen time series in [11]. Although it was originally introduced as a model for purely TSF problems, the authors showed how the architecture resembles a meta-learner that adapts weights used for the final prediction. Nevertheless, the model itself does not provide explicitly a way to fine-tune using some samples (few-shot learning).

To deal with Time Series Classification (TSC) in few-shot settings, Narwariya et al. [9] show how MAML can be adapted with minor modifications. They achieve better results compared to the baselines, where the Resnet achieved the closest performance results. Besides MAML, attention mechanisms have been used lately as an approach for leveraging the support set on text classification [7] and time series forecasting [6].

3   Multimodal Meta-Learning for TSR

3.1   Problem Definition

We define generally meta-learning for time series regression as, given an ordered set $\mathcal{D}_j = \{(\mathbf{x}_n, y_n)\}_{n=1:N}$, the problem of learning a method that adapts the parameters θ of a regression model f_θ by using just $\mathcal{D}_j^s = \{(\mathbf{x}_n, y_n)\}_{n=1:Q}$ such that it performs well on $\mathcal{D}_j^q = \{(\mathbf{x}_n, y_n)\}_{n=Q:N}$. Where $\mathbf{x}_n \in \mathrm{R}^{L \times C}$ denotes a time window with labels $y_n \in \mathrm{R}$, whereas $\mathcal{D}_j^s$ and $\mathcal{D}_j^q$ are the support and query sets, respectively. The union of both sets, $\mathcal{D}_j = \{(\mathbf{x}_n, y_n)\}_{n=1:N}$, is a time series regression (TSR) task, a definition inspired by the concept of task in meta-learning applied to image classification [3,19]. Every fixed-length window is intended to be the input for the regressor.

Formally, the optimization objective can be defined as

$$\min_{\phi, \theta} \sum_j \mathcal{L}_j(\mathcal{D}_j^q, f_{\theta^*}) \tag{1}$$

where $\theta^* = \mathcal{U}(\theta, \mathcal{D}_j^s, \phi)$ is a learner or an update rule for the parameters θ that depends on the support set $\mathcal{D}_j^s$ and the meta-parameters ϕ. The task loss $\mathcal{L}_j$

measures the performance of the predicted labels accounting for all the labeled windows $\mathcal{L}_j(\mathcal{D}_j^q, f_{\theta|\phi}) = \sum_n \mathcal{L}(y_n, f_\theta(\mathbf{x}_n))$. Since this is a regression problem, the loss can be, for instance, the Mean Squared Error (MSE) or the Mean Absolute Error (MAE).

When applied to a new TSR task, the learner uses the update rule to estimate a new set of parameters θ^* by just using Q samples (or time windows in this context). Q is typically small, therefore training a network from scratch with $\mathcal{D}_j^s$ is not possible without overfitting. However, in order to tackle this problem, there is normally a small set of long multivariate time series (including the target channel) $\mathcal{S}$ available for training such that $\mathcal{S} = \{(\mathbf{S}_i, Y_i)|\mathbf{S}_i \in \mathrm{R}^{L_i \times C}, Y_i \in \mathrm{R}^{L_i}, i = 1, ..., M\}$, where $\mathbf{S}_i$ denotes the input channels and Y_i, the output or target channel to be predicted.

3.2   Meta-Windows: Redesigning Tasks for TSR

Independently from the approach to solve the problem formulated in Eq. 1, it is necessary to have a lot of tasks available for training, this is also the case for meta-learning applied to image classification. Nevertheless, the situation is usually to have long but few multivariate time series. In this subsection, we introduce a redesign of this setting to overcome this challenge.

Long time series are difficult to feed into a model, therefore the common approach consists in creating smaller fixed-length windows through a windows generation process that uses a rolling window, $\mathcal{W}_{\delta,k}(\cdot)$, where δ denotes the window size and k is the step size for the windows generation. Given a long multivariate time series with target channel, $(\mathbf{S}, Y) \in \mathcal{S}$, we generate the set of labeled windows $\mathcal{D}$ such that $\mathcal{D} = \mathcal{W}_{\delta,k}(\mathbf{S}, Y) = \{(\mathbf{x}_n, y_n), \mathbf{x}_n = \mathbf{S}_{(j\cdot k):(j\cdot k+\delta)}, y_n = Y_{(j\cdot k+\delta)}, j = 1, ..., L - \delta\}$, where L is the length of the multivariate time series and the lower indexing on S and Y refers to the time axis of the time series.

The methods introduced in this paper leverage these windows by grouping them in **meta-windows**. All meta-windows contain the same number of labeled windows (denoted as l), whereas every window belongs to only one window. The Algorithm 1 explains how the meta-windows are generated from long time series.

If the time series are not periodic, it is possible to assume that the windows are very correlated to other temporally close windows, but are less correlated to the temporally far samples. In fact, this can be supported by looking at the monotonically decreasing auto-correlation diagram of Y_i (see supplementary material[1]). This results in a meta-window $\mathcal{T}_t \in \mathcal{T}$ that is correlated with its neighbor $\mathcal{T}_{t+1}$, but approximately uncorrelated with other meta-windows.

Based on the above-mentioned assumption on temporally correlation, we redesign a task for TSR such that two continuous windows are considered belonging to the same task. It means, after the defined problem in Eq. 1, we can generate a lot of tasks by setting a sampled meta-window $\mathcal{T}_t \sim \mathcal{T}$ as the support set $\mathcal{D}_j^s$ and the next one $\mathcal{T}_{t+1}$ as the query set. Also, due to the decreased correlation

---

[1] Accessible in https://www.dropbox.com/s/tuzs6l8zy9zyon9/AALTD_21_MMAML_TSR_Supplementary.pdf?dl=0.

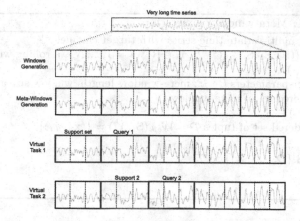

Fig. 1. Task design for a univariate time series, also applicable to multivariate time series. We omit the other channels and the respective target associated to every window for the sake of simplicity.

with temporally-far meta-windows, this design guarantees certain level of diversity. Moreover, including meta-windows generated from different long time series may increase this task diversity. Since the generated pair of sets support-query aim to "simulate" a new TSR task, we refer to them as *virtual tasks*. The Fig. 1 illustrates this procedure.

3.3   MAML for TSR

In this section, we formally define our proposed algorithm MAML for TSR. We denote $\mathcal{T}_t$ as the t-th meta-window, sampled from a distribution $p(\mathcal{T})$. We assume that the meta-windows were generated from a set of long multivariate time series coming from different instances but with the same semantics among the channels (i.e. three accelerometer measurements from different subjects).

Given that the meta-windows, coming from the same long time series, are temporally ordered through a temporal index t, we want to use the meta-window $\mathcal{T}_t$ as the support set, while setting the subsequent meta-window $\mathcal{T}_{t+1}$ as the query. The idea of MAML for TSR is, then, to find an initial set of parameters θ^* such that a gradient descent optimizer adapts the model f_θ in one step (or few steps) to a new domain by using just a meta-window $\mathcal{T}_t$. This optimization objective can be expressed formally as:

$$\min_\theta \sum_{\mathcal{T}_t \sim p(\mathcal{T})} \mathcal{L}_{\mathcal{T}_{t+1}} \left(f_{\theta - \alpha \nabla_\theta \mathcal{L}_{\mathcal{T}_t}(f_\theta)} \right)$$

$$= \min_\theta \sum_{\mathcal{T}_t \sim p(\mathcal{T})} \sum_{(\mathbf{x}_j, y_j) \in \mathcal{T}_{t+1}} ||y_j - f_{\theta - \alpha \nabla_\theta \mathcal{L}_{\mathcal{T}_t}(f_\theta)}(\mathbf{x}_j)||_1 \tag{2}$$

Algorithm 1: Meta-windows generation.

Input: Long multi-variate time series with target channel
$\mathcal{S} = \{(\mathbf{S}_i, Y_i) | \mathbf{S}_i \in \mathrm{R}^{L_i \times C}, Y_i \in \mathrm{R}^{L_i}, i = 1, ..., N\}$, a rolling window
$\mathcal{W}_{\delta,k}(\cdot)$
Input: Hyper-parameters l, δ, k : meta-window length, window size and step size
1 Initialize ordered set of meta-windows $\mathcal{T}$
2 **for all** $(\mathbf{S}_i, Y_i) \in \mathcal{S}$:
3   Create ordered set of tuples $\mathcal{D} = \mathcal{W}_{\delta,k}(\mathbf{S}_i, Y_i) = \{(\mathbf{x}_n, y_n)\}$
4   **for** $n = 1, ..., \lfloor \frac{|\mathcal{D}|}{l} \rfloor$:
5   Append $\{\mathcal{D}_{n \cdot l : (n+1) \cdot l}\}$ to $\mathcal{T}$
6   **end for**
7 **end for**

---

which is based on the formulation of MAML by Finn et al. [3] for fast adaptation in classification with neural networks. The second line reformulates the loss by using MAE as the task loss $\mathcal{L}_{\mathcal{T}}$. In the Algorithm 2, we detail the process for optimizing the proposed loss function in Eq. 2.

3.4 MMAML for TSR

In this section, we introduce the multi-modal model-agnostic meta-learning (MMAML) for TSR which draws inspiration from [22]. This approach takes a **modulation network** that changes the parameters of a **task network** which makes the final prediction. The parameters are modulated according to meta-task information that is extracted by the modulation network. Therefore, the modulation network is a feature extractor at the task level (or meta-window level), whereas the task network processes single windows. The extracted information from the meta-windows is useful for the fast adaptation. In the supplementary material, we show that the embeddings of the meta-windows are forming groups with the other ones coming form the same long time series.

Encoding task information means in our current work to embed the support set, which is a meta-window. Vuorio et al. [22] use relational networks to process images and targets belonging to the support set. However, as we are interested in embedding meta-windows (time series), we need to perform the task encoding differently.

Meta-Windows Encodings. As explained before, the support set (task information) is a meta-window $\mathcal{T} = \{(\mathbf{x}_i, y_i), i = 1, .., l\}$ and is originated from a long time series $(\mathbf{S}, Y)$. As a way of simplifying the meta-window, so that it can be input to the modulation network without redundant information and by avoiding a huge overhead in training, we propose to summarize it by concatenating the first sample of every window (and every channel) belonging to the meta-window. An additional channel is created after concatenating similarly the respective target signal. Therefore, the summarization is performing a downsampling of the

---

Algorithm 2: MAML for TSR

Input: $p(\mathcal{T})$: distribution over meta-windows, with indexed, and temporally
ordered meta-windows $\mathcal{T}_1, \mathcal{T}_2, ...$
Input: α: learning rate, β: meta-learning rate, $\mathcal{L}_{\mathcal{T}}$: task loss

1 randomly initialize θ
2 **while** not done **do**
3   Sample batch of meta-windows $\mathcal{T} \sim p(\mathcal{T})$
4   **for all** $\mathcal{T}_t \in \mathcal{T}$ **do:**
5   Set windows $\mathcal{T}_t = \left\{ \mathbf{x}_i^{(t)}, y_i^{(t)} \right\}$
6   Evaluate $\nabla_\theta \mathcal{L}_{\mathcal{T}_t} (f_\theta)$ using $\mathcal{T}_t$ and $\mathcal{L}_{\mathcal{T}_t}$
7   Compute parameter updates : $\theta_t' = \theta - \alpha \nabla_\theta \mathcal{L}_{\mathcal{T}_t} (f_\theta)$
8   Save windows $\mathcal{T}_{t+1} = \left\{ \mathbf{x}_j^{(t+1)}, y_j^{(t+1)} \right\}$ for meta-update
9   **end for**
10   Update $\theta \leftarrow \theta - \beta \nabla_\theta \sum_{\mathcal{T}_t \in \mathcal{T}} \mathcal{L}_{\mathcal{T}_{t+1}} \left(f_{\theta_t'} \right)$ using dataset $\mathcal{T}_{t+1}$ and a given
model loss $\mathcal{L}_{\mathcal{T}_{t+1}}$.
11 **end while**

---

meta-window. After summarizing the support set, we obtain a multivariate-time
series (MTS), which can be encoded through any representation learning algorithm. For learning this latent representation, we use a variational recurrent
auto-encoder (VRAE) [2].

Modulation and Task Network. We introduce the modulation network to
be applied in our work, which comprises three sub-modules:

- **Encoder.** It embeds the input (summarized meta-window $\mathcal{T}'$) to a latent
  representation, such that $z = h_{\theta_{enc}}(\mathcal{T}')$ using variational bayes.
- **Decoder.** It reconstructs the input, formally $\hat{\mathcal{T}} = h_{\theta_{dec}}(z)$, aiming to mini-
  mize the reconstruction loss $\|\mathcal{T}' - \hat{\mathcal{T}}\|^2$.
- **Generator.** It outputs the parameters ρ that modify the parameters θ from
  the main network through FiLM layers [14]. Formally, $\rho = h_{\theta_{gen}}(z)$.

The task network is composed of a feature extractor $\phi_{\theta_{ext}}$ and a last (linear)
layer θ. The final output for an input $\mathbf{x}$, given the output of the generator ρ, is
then computed as:

$$\hat{y} = f_{\theta|\rho}(\mathbf{x}) = FiLM(\theta|\rho)^T \phi_{\theta_{ext}}(\mathbf{x}) \tag{3}$$

Here we have a set of four types of parameters $\omega = \{\theta, \theta_{dec}, \theta_{enc}, \theta_{gen}, \theta_{ext}\}$, but
θ being the most interesting ones as they are the only fine-tuned parameters for
fast adaptation in this work. In general, however, the parameters of the feature
extractor can be also included in the adaptation.

Algorithm 3: MMAML for TSR

Input: $p(\mathcal{T})$: distribution over meta-windows, with indexed, and temporally ordered meta-windows $\mathcal{T}_1, \mathcal{T}_2, \ldots$

Input: α, β : step size hyper-parameters, λ : weight for the variational loss

1 randomly initialize $\omega = \{\theta, \theta_{dec}, \theta_{enc}, \theta_{gen}, \theta_{ext}\}$
2 **while** not done **do**
3   Sample batch of meta-windows $\mathcal{T} \sim p(\mathcal{T})$
4   **for all** $\mathcal{T}_t \in \mathcal{T}$ **do:**
5   Set windows $\mathcal{T}_t = \{\mathbf{x}_i, y_i\}$
6   Get meta-window summary: $\mathcal{T}'_t = \text{Summarize}(\mathcal{T}_t)$
7   Infer embedding: $z = h_{\theta_{enc}}(\mathcal{T}'_t)$
8   Generate parameters: $\rho = h_{\theta_{gen}}(z)$
9   Modulate last layer parameters: $\hat{\theta} = \text{FiLM}(\theta|\rho)$
10   Evaluate $\nabla_\theta \mathcal{L}_{\mathcal{T}_t}(f_{\hat{\theta}})$ using $\mathcal{T}_t$ and $\mathcal{L}_{\mathcal{T}_t}$
11   Compute parameter updates: $\theta'_t = \theta - \alpha \nabla_\theta \mathcal{L}_{\mathcal{T}_t}(f_{\hat{\theta}})$
12   Save meta-window $\mathcal{T}_{t+1} = \{\mathbf{x}_j, y_j\}$ for meta-update
13   **end for**
14   Reconstruct meta-windows $\mathcal{T}_t \in \mathcal{T}: \hat{\mathcal{T}}_t = h_{\theta_{dec}}(h_{\theta_{enc}}(\mathcal{T}'_t))$
15   Update $\omega \leftarrow \omega - \beta \nabla_\omega \sum_{\mathcal{T}_t \in \mathcal{T}} \left(\mathcal{L}_{\mathcal{T}_{t+1}} \left(f_{\theta'_t|\rho} \right) + \lambda \mathcal{L}_{VAE}(\mathcal{T}'_t) \right)$
16 **end while**

The loss for MAML (Eq. 2) can be extended easily to a formulation that includes the modulation network.

$$\sum_{(\mathbf{x}_j, y_j) \in \mathcal{T}_{t+1}} ||y_j - f_{\theta - \alpha \nabla_\theta \mathcal{L}_{\mathcal{T}_t}(f_{\theta|\rho})|\rho}(\mathbf{x}_j)||_1 + \lambda \cdot \mathcal{L}_{VAE}(\mathcal{T}'_t) \quad (4)$$

We have included the variational loss that can be re-phrased in our context, by denoting $\mathcal{T}'$ as the meta-window summary, in the following way:

$$\mathcal{L}_{VAE}(\mathcal{T}'_t) = ||\mathcal{T}'_t - \hat{\mathcal{T}}_t||^2 + \mathcal{KL}[\mathcal{N}(\mu(\mathcal{T}'_t), \Sigma(\mathcal{T}'_t))||\mathcal{N}(0, I)], \quad (5)$$

where $\mathcal{KL}$ is the Kullback-Leiber divergence, I is an identity matrix and $\mu(\cdot), \Sigma(\cdot)$ are mean and covariance functions respectively, which are modelled by the encoder.

The Algorithm 3 introduces MMAML for TSR. It comprises an outer loop which modifies all the involved parameters ω, and an inner-loop that just involves the parameters meant to be updated during the fast adaptation, θ.

The Fig. 2 illustrates the different modules. Note that there are two networks: the modulation and task network. The decoder is considered outside the task network as it is not used in inference time. More importantly, the input of the task network is just a window, while the input for the modulation network is a summarized meta-window.

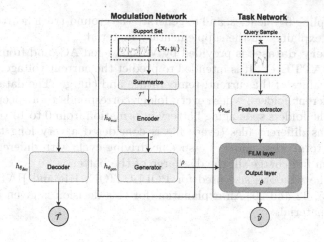

Fig. 2. MMAML for TSR architecture (based on [22]).

4 Experiments

In this section, we test our proposed ideas empirically to assess how much improvement meta-learning can bring compared to other approaches. Furthermore, we are interested in analyzing the performance of the models after several gradient steps and in long-term, in other words, in time horizons very far from the support meta-window that has been used for the adaptation.

4.1 Datasets

We perform the experiments on three different datasets which belong to different use cases.

The **air pollution dataset** (whose original name is *2.5 Data of Chinese Cities Dataset*[2], and from now on referred as **POLLUTION**) contains the PM 2.5 data from Beijing, Shanghai, Guangzhou, Chengdu and Shenyang, which includes as input channels meteoreological variables, whereas the target channel corresponds to the PM2.5 particles concentrations. The time period of the data spans six years, between Jan 1st, 2010 to Dec. 31st, 2015. In this dataset, every city corresponds to a very long time series that is the source for the meta-window generation.

The **heart rate dataset** (whose original name is *PPG-DaLiA Dataset*[3], from now on referred as **HR**) is to be used for PPG-based heart rate estimation. It includes physiological and motion variables, recorded from wrist- and chest-worn devices, on 15 subjects performing different activities, under real-life conditions. We consider every subject as a very-long time series. A careful pre-processing is necessary to synchronize the variables samples, as they have

---

[2] https://archive.ics.uci.edu/ml/datasets/PM2.5+Data+of+Five+Chinese+Cities.
[3] https://archive.ics.uci.edu/ml/datasets/PPG-DaLiA.

different sampling frequencies, and to generate the ground-truth heart rate. More about this is explained in the supplementary material.

The **battery dataset** is provided by Volkgswagen AG, and from now on is referred as **BATTERY**. It is intended to predict the current voltage, given past and current values of the current, temperature and charge. The data is divided in eleven different folders, where every folder corresponds to a specific battery age. Hence, the folders span aging battery information from 0 to 10 years. Every folder contains different files (every file is considered a very long time series, having 96 in total) that matches a specific driving cycle with different patterns of speed, from full charge till the discharge of the battery.

Finally, the windows size used for POLLUTION, HR and BATTERY are respectively 5, 32 and 20. An explanation for this decision is given in the supplementary material.

4.2 Baselines

We compare our proposed methods with the following models:

- **Target Mean** We predict for all the query samples the mean of the target from the support set such that $\hat{y} = \frac{1}{l} \sum_{(\mathbf{x}_i, y_i) \in \mathcal{T}_t} y_i$. However, this corresponds to an unreal scenario, as we are assuming that the target channel is not available in inference time, but just in fast adaptation.

- **Resnet** We used the implementation of [21] keeping the same architecture (filter size and number of filters). The number of parameters of the final model is about 500.000. This model is considered, because it has shown competitive results in time series regression [21]. Before adapting the model to a new (virtual) task, we pre-train the network on the raw meta-training dataset using standard training. Then, we apply transfer learning, where we freeze the whole network parameters but the last layer (128 parameters).

- **VRADA** It [15] combines two networks: a VRNN as proposed by [1] and DANN [4], which includes a domain classifier and a regressor. The VRNN is a one-layer LSTM, with hidden dimension and the latent dimension equal to 100 (as in the original paper). The domain classifier comprises two fully connected layers with 100 and 50 neurons respectively, with output layer equal to the *number of different classes* (which in this case we assume are each of the original long time series in $\mathcal{S}$), whereas the label predictor is a regressor with similar architecture but with output layer equal to 1. The hidden layers use DropOut (drop-out probability = 0.5), Batch Normalization and ReLU. The number of total parameters for VRADA is around 236.000. It is considered as a baseline since its architecture allows to extract domain-invariant features.

- **LSTM** [5] We applied an architecture with 120 neurons, 2 layers and a linear output layer. The total number of parameters, varying according to every dataset, is around 180.000. Although this is also the model used for applying MAML and MMAML, we use as baseline a standard fine-tuning, after pre-training a network without meta-learning.

4.3   Experimental Setup

Given a set of few very long-time series, we split them in three disjunctive sets: training, validation and testing. 60% of the very-long time series are assigned to training, 20% to validation and 20% to test. For every split, we run a rolling window that generates a set of labeled windows. Subsequently, meta-windows are generated by applying the Algorithm 1 on every split, thus originating the datasets splits used on meta-learning: meta-training, meta-validation and meta-testing respectively[4].

MAML and MMAML use the meta-training and meta-validation datasets, whereas the baselines are trained using a training and validation datasets. After training (or meta-training), the final parameters are available to be used as initialization for fine-tuning. How good the models after fine-tuning are is evaluated using a meta-testing protocol, where every experiment is run five times to report the mean and the 95% confidence interval.

The meta-testing follows the same procedure for the baselines and MAML/MMAML. We draw a meta-window as the support set, and use it to adapt (or fine-tune) the model, while the following meta-windows (up to a given horizon H) are used as query set. By doing this iteratively, many virtual tasks are available for the meta-testing, as simply sliding over the ordered set of meta-windows with a given step size (meta-testing step size) gives different virtual tasks. For instance, Fig. 3 depicts two virtual tasks separated by a meta-learning step size equals to two. At the end, the total error is computed as the average error over all the queries. To follow the proposed meta-testing procedure, two parameters must be provided: the meta-testing step size and horizon.

Specifically, we experiment fine-tuning with 1 and 10 gradient updates of the **last layer** parameters, except for VRADA, where all the parameters of its regressor module are updated. For the baselines, we also experimented with updates including weight decay. Note that MAML and MMAML also consider this number of updates in the inner-loop during meta-training, not only in the meta-testing. Additionally, in order to assess how good was the adaptation to the new virtual task on the long term, we evaluate on a query set that includes then 10 following meta-windows (10 horizons). The meta-testing step-size is set to $\lfloor \frac{M}{100} \rfloor$, where M is the number of meta-windows generated from the long time series $\mathcal{S}_i = (\mathbf{S}_i, Y_i) \in \mathcal{S}$. Additionally, we experiment with a meta-window length equal to 50 ($l = 50$).

The training phase of the baselines uses the training set for optimizing the parameters of the whole network by using mean absolute error (MAE) as loss $\mathcal{L}_{T_t}$ and the labeled windows. The number of iterations was limited by the early stopping criteria, and using the error in the validation/meta-validation set as a reference. If the error in the validation set does not decrease within a given number of iterations (so-called **patience**), then the training stops. A **training**

---

[4] We provide the created meta-windows for the splits of POLLUTION and HR as pickled numpy objects in https://www.dropbox.com/sh/yds6v1uok3bjydn/AAC5GRWw0F3clopRlk00Smvza?dl=0.

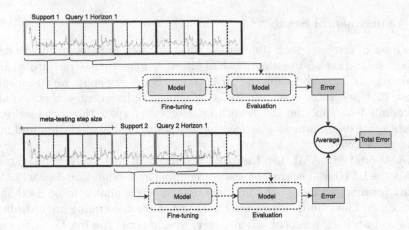

Fig. 3. Meta-testing Protocol. This procedure enables a lot of tasks during evaluation.

epoch is considered to be the group of updates after passing over the whole training set.

Similarly, the training of MAML/MMAML, here referenced as meta-training, follows the Algorithm 2 and 3 respectively, only adapting the last layer parameters. Early stopping was also applied, so that it stops when the meta-validation error starts to increase after some epochs. In this context, a **meta-training epoch** is one outer-loop iteration in the Algorithm 3, therefore it does not compute updates for the whole set of meta-windows during one epoch, since only a group of them are sampled.

For the baselines, we set the following parameters for all the experiments: batch size is 128, training epochs are 1000, and the patience for the early stopping is 50. In the experiments with baselines, two learning rates are considered. On the one side, the training learning rate that is used for finding the pre-trained parameters (*Training LR*). On the other side, a different learning rate is considered for fine-tuning the models during test (*Fine-tuning LR*). Both learning rates are chosen from a set of possibilities $\{0.01, 0.001, 0.0001\}$, by assessing the performance of the model with the pre-trained parameters and the performance after fine-tuning accordingly on the validation set. For VRADA, the training LR is fixed to 0.0003, following the value used by [15]. The weight decay for the fine-tuning is chosen from $\{0, 0.5, 0.1, 0.01, 0.001, 0.0001\}$.

MAML algorithm uses the same LSTM architecture as the baseline, and aims to find the best last-layers parameters so that it achieves a good performance in few updates for a new virtual task. It uses the following settings. The batch size (batch of meta-windows in Algorithms 2) is 20. The meta-training epochs are set to 10000, similar to the baselines. The patience is 500, as the definition of meta-training epoch is slightly different to training epoch, and less data is considered in every epoch. We also include a noise level, to achieve meta-augmentation [16] by adding noise to the targets of the support sets, such that $y_{noise} = y + \epsilon, \epsilon \sim \{0, noise\ level\}$. This is one of the proposed approaches [16] to make the model more robust against meta-overfitting. The grid for the hyper-parameter tuning via the meta-validation set are set as follows: meta-learning rate (β in Algorithm 2) $\{0.0005, 0.00005\}$, learning rate (α) $\{0.01, 0.001, 0.0001\}$ and noise level $\{0, 0.01, 0.001\}$. The hyperparameters are chosen so that it reduces the error for horizon 10 in all the experiments with MAML.

MMAML uses the same task network as MAML, a two layers LSTM, how-ever the architecture involves other modules. The encoder and the decoder are one-layer LSTMs with hidden size 128. The generator is a linear layer with 256 neurons, as it generates two vectors of parameters for the FiLM layer (each of dimensionality 128). We tune the same hyper-parameters as for MAML, but including the VRAE weight (λ in Eq. 4), considering the grid $\{0.1, 0.001, 0.0001\}$. The final chosen configurations for the baselines and the proposed models are presented in the support material.

4.4   Results

We present the results of the proposed experiments in the Table 1. The bold font indicates the best results (lowest MAE) and the underlined font indicates the second best result. Additionally, the 95% confidence interval is provided.

After the results, it is possible to observe that MMAML achieves overall good results, always better than transfer-learning on the same backbone (LSTM) and, most of the times, better than transfer learning even on more powerful models such as Resnet. Moreover, the performance, after fine-tuning, remains high for a long evaluation horizon when using MAML and MMAML as can be seen in the results on 10 horizons. It means they exhibit less temporal overfitting.

Another important insight is that MAML is less prone to overfitting after more gradient steps. By looking at the results on all the datasets, it is noticeable that on 10 gradient steps, the MAML algorithm still performs better in long horizons (10 Horizons) than MMAML. We hypothesize that this is because the parameter modulation from MMAML based on meta-features of a meta-window is more robust against overfitting so long as it is applied on meta-windows tem-porally close to the support meta-window (input of the modulation network), but might decrease the performance in temporally-far examples.

During the baselines evaluation, we notice that finding the best *fine-tuning LR* given a pre-trained model is difficult, since the learning rate that works best for the validation set may not perform equally good for the meta-testing. They may overfit easily, as it happened on HR (1 Horizon, 1 gradient step), where the simplest baseline performed better. Here, our proposed methods set a clear advantage as the *fine-tuning LR* is already fixed in the meta-training process. However, if the validation and test set are somewhat similar, for instance due to a small domain shift, the *fine-tuning LR* tuned on the validation set may be suitable enough for the test set. Thus, a more powerful model will have an advantage over meta-learning approaches, as the results on BATTERY show.

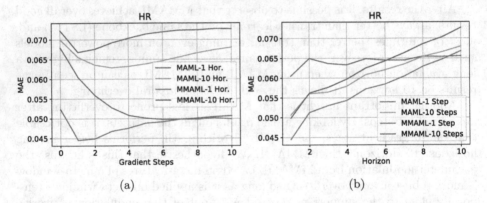

Table 1. Results

Dataset	Model	1 gradient step		10 gradient steps	
		1 Horizon	10 Horizons	1 Horizon	10 Horizons
POLLUTION	Target Mean	0.0465 ± 0.0000	0.0495 ± 0.0000	0.0465 ± 0.0000	0.0495 ± 0.0000
	Resnet	0.0491 ± 0.0074	0.0502 ± 0.0062	0.0472 ± 0.0047	0.0519 ± 0.0057
	VRADA	0.0444 ± 0.0012	0.0428 ± 0.0011	0.0438 ± 0.0008	0.0429 ± 0.0008
	LSTM	0.0467 ± 0.0009	0.0463 ± 0.0012	0.0446 ± 0.0006	0.0437 ± 0.0005
	MAML (ours)	0.0421 ± 0.0002	0.0418 ± 0.0003	0.0423 ± 0.0010	**0.0416 ± 0.0009**
	MMAML (ours)	**0.0410 ± 0.0012**	**0.0417 ± 0.0007**	**0.0411 ± 0.0010**	0.0420 ± 0.0011
HR	Target Mean	0.0542 ± 0.0000	0.0975 ± 0.0000	0.0542 ± 0.0000	0.0975 ± 0.0000
	Resnet	0.0670 ± 0.0063	0.0817 ± 0.0035	0.0625 ± 0.0043	0.0734 ± 0.0024
	VRADA	0.0789 ± 0.0066	0.0799 ± 0.0060	0.0761 ± 0.0084	0.1140 ± 0.0071
	LSTM	0.0673 ± 0.0002	0.0788 ± 0.0006	0.0565 ± 0.0004	0.0906 ± 0.0010
	MAML (ours)	0.0634 ± 0.0018	0.0792 ± 0.0029	0.0511 ± 0.0022	**0.0711 ± 0.0068**
	MMAML (ours)	**0.0448 ± 0.0009**	**0.0689 ± 0.0015**	**0.0507 ± 0.0008**	0.0729 ± 0.0027
BATTERY	Target Mean	0.0255 ± 0.0000	0.0658 ± 0.0000	0.0255 ± 0.0000	0.0658 ± 0.0000
	Resnet	**0.0184 ± 0.0024**	**0.0141 ± 0.0007**	**0.0091 ± 0.0007**	0.0160 ± 0.0012
	VRADA	0.0352 ± 0.0019	0.0309 ± 0.0016	0.0967 ± 0.0135	0.0995 ± 0.0124
	LSTM	0.0407 ± 0.0025	0.0417 ± 0.0024	0.0195 ± 0.0013	0.0217 ± 0.0013
	MAML (ours)	0.0243 ± 0.0012	0.0170 ± 0.0010	0.0135 ± 0.0006	**0.0115 ± 0.0004**
	MMAML (ours)	0.0206 ± 0.0021	0.0154 ± 0.0020	0.0156 ± 0.0023	0.0149 ± 0.0022

Fig. 4. Change in MAE while increasing the gradient steps during fine-tuning on meta-testing.

4.5 Ablation Studies

We run some additional experiments to test the performance of our proposed algorithms under different configurations. Firstly, we would like to see how the error behaves when the models are fine-tuned beyond the number of gradient steps assumed during the meta-training. A look at Fig. 4 makes possible to understand this and how the overfitting may arise on our proposed methods under different horizons. The lowest error is achieved after one gradient step even on long horizons, except for MAML on 1 horizon, where MAE keeps decreasing after several gradient updates (Fig. 4a). This shows that MAML is robust against overfitting on close horizons, after several gradient steps. However, when having more updates, MAML may overfit temporally, thus exhibiting bad performance in long horizons as the orange curve depicts in Fig. 4b.

The Figs. 5a, 5b and 5c show that there is indeed an advantage of including the variaional loss in our formulation, as there is a decreased MAE when having values for the VRAE weight (λ) different from zero.

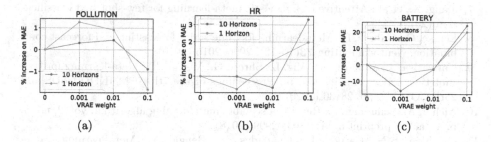

Fig. 5. Change in MAE with respect to VRAE.

5 Conclusion

The present work introduces an extension of Meta-Agnostic Meta-Learning (MAML) and Multi-modal MAML (MMAML) to time series regression (TSR). We propose a design for the tasks such that we leverage the original, more common scenario of few but long time series available. The proposed design, which introduces the concept of "meta-window", makes possible to have more tasks available for meta-training. Through experiments, we show how this idea works on different datasets, allowing to achieve better performance than traditional methods such as transfer learning. This is the first time to apply meta-learning for fast adaption on TSR, and it shows that it is possible to adapt to new TSR tasks with few data and within few iterations. For future work, we hypothesize that the application of the introduced ideas would have promising results in time series forecasting.

Acknowledgements. The research of Kiran Madhusudhanan is co-funded by the industry project "IIP-Ecosphere: Next Level Ecosphere for Intelligent Industrial Production". Sebastian Pineda Arango would also like to thank Volkswagen AG who funded his internship in order to carry out this research.

References

1. Chung, J., Kastner, K., Dinh, L., Goel, K., Courville, A.C., Bengio, Y.: A recurrent latent variable model for sequential data. In: Advances in Neural Information Processing Systems, pp. 2980–2988 (2015)
2. Fabius, O., van Amersfoort, J.R.: Variational recurrent auto-encoders. arXiv preprint arXiv:1412.6581 (2014)

3. Finn, C., Abbeel, P., Levine, S.: Model-agnostic meta-learning for fast adaptation of deep networks. arXiv preprint arXiv:1703.03400 (2017)
4. Ganin, Y., et al.: Domain-adversarial training of neural networks. J. Mach. Learn. Res. **17**(1), 2030–2096 (2016)
5. Hochreiter, S., Schmidhuber, J.: Long short-term memory. Neural Comput. **9**(8), 1735–1780 (1997)
6. Iwata, T., Kumagai, A.: Few-shot learning for time-series forecasting. arXiv preprint arXiv:2009.14379 (2020)
7. Jiang, X., et al.: Attentive task-agnostic meta-learning for few-shot text classification (2019)
8. Lemke, C., Gabrys, B.: Meta-learning for time series forecasting and forecast combination. Neurocomputing **73**(10–12), 2006–2016 (2010)
9. Narwariya, J., Malhotra, P., Vig, L., Shroff, G., Vishnu, T.: Meta-learning for few-shot time series classification. In: Proceedings of the 7th ACM IKDD CoDS and 25th COMAD, pp. 28–36 (2020)
10. Nichol, A., Schulman, J.: Reptile: a scalable metalearning algorithm, vol. 2, no. 3, p. 4 . arXiv preprint arXiv:1803.02999 (2018)
11. Oreshkin, B.N., Carpov, D., Chapados, N., Bengio, Y.: Meta-learning framework with applications to zero-shot time-series forecasting. arXiv preprint arXiv:2002.02887 (2020)
12. Oreshkin, B.N., Carpov, D., Chapados, N., Bengio, Y.: N-BEATS: neural basis expansion analysis for interpretable time series forecasting. In: International Conference on Learning Representations (2020)
13. Oreshkin, B.N., López, P.R., Lacoste, A.: TADAM: task dependent adaptive metric for improved few-shot learning. In: NeurIPS, pp. 719–729 (2018)
14. Perez, E., Strub, F., de Vries, H., Dumoulin, V., Courville, A.C.: Film: visual reasoning with a general conditioning layer. In: AAAI, pp. 3942–3951 (2018)
15. Purushotham, S., Carvalho, W., Nilanon, T., Liu, Y.: Variational recurrent adversarial deep domain adaptation. In: ICLR (2017)
16. Rajendran, J., Irpan, A., Jang, E.: Meta-learning requires meta-augmentation. In: Advances in Neural Information Processing Systems, vol. 33 (2020)
17. Ravi, S., Larochelle, H.: Optimization as a model for few-shot learning. In: International Conference on Learning Representations (ICLR) (2017)
18. Santoro, A., Bartunov, S., Botvinick, M., Wierstra, D., Lillicrap, T.: Meta-learning with memory-augmented neural networks. In: International Conference on Machine Learning, pp. 1842–1850 (2016)
19. Snell, J., Swersky, K., Zemel, R.: Prototypical networks for few-shot learning. In: Advances in Neural Information Processing Systems, pp. 4077–4087 (2017)
20. Talagala, T.S., Hyndman, R.J., Athanasopoulos, G., et al.: Meta-learning how to forecast time series. Monash Econometrics Bus. Stat. Working Pap. **6**, 18 (2018)
21. Tan, C.W., Bergmeir, C., Petitjean, F., Webb, G.I.: Time series extrinsic regression. arXiv preprint arXiv:2006.12672 (2020)
22. Vuorio, R., Sun, S.H., Hu, H., Lim, J.J.: Multimodal model-agnostic meta-learning via task-aware modulation. In: Advances in Neural Information Processing Systems, pp. 1–12 (2019)

Cluster-Based Forecasting for Intermittent and Non-intermittent Time Series

Tom van de Looij[1]([✉]) and Mozhdeh Ariannezhad[2]

[1] University of Amsterdam, Amsterdam, The Netherlands
tom.vandelooij@student.uva.nl
[2] AIRLab, University of Amsterdam, Amsterdam, The Netherlands
m.ariannezhad@uva.nl

Abstract. Producing accurate forecasts is an essential part of successful inventory management for any retail business. Previous research has shown that the clustering of time series data into disjoint clusters can reduce the forecast error, eventually leading to cost savings. A common measure used to cluster time series data is Dynamic Time Warping. While it can handle time series of different length and guarantees to provide the optimal alignment, it is computationally expensive and assumes that one time series is a stretched non-linear version of another time series. For datasets containing intermittent time series, i.e. showing no clear structure, DTW is not the best suited method. In this paper, we propose a new framework that uses Simple Exponential Smoothing (SES) and a Self-Organizing Map (SOM) that is able to improve the clustering performance for clusters containing intermittent and non-intermittent time series. Using LightGBM as the forecasting model, we evaluate our approach on a real-world dataset, and find that the computational time can be reduced substantially compared to DTW when using a combination of SOM and LightGBM for both intermittent and non-intermittent time series while maintaining similar levels of accuracy.

Keywords: Intermittent time series · Clustering · Forecasting · Self-organizing map · Hierarchical agglomerative clustering · Dynamic time warping

1 Introduction

For any business, sales forecasting is a critical task in order to maintain a correct inventory level, where purchasing too few units of a product leads to a lost sales opportunity and buying too much units results in an overstock. Therefore, producing accurate forecasts is an essential component in maintaining an efficient supply chain [21]. According to [26], intermittent demand may constitute up to 60% of the total stock value. As a result, small improvements in the management of intermittent stock items can contribute to substantial cost savings. Intermittent time series are characterized by multiple non-demand intervals and do not

© Springer Nature Switzerland AG 2021
V. Lemaire et al. (Eds.): AALTD 2021, LNAI 13114, pp. 139–154, 2021.
https://doi.org/10.1007/978-3-030-91445-5_9

contain enough data to model trend and seasonality [19]. This makes intermittent time series hard to forecast as there are two sources of uncertainty: the sporadic nature of the demand level and the timing of the demand occurrence [17].

Over the last five decades, since the first specialised method for intermittent demand was proposed in [5], there have been very few new forecasting methods specifically designed for intermittent time series compared to fast-moving time series [17]. Previous research has shown that partitioning the dataset into disjoint clusters can reduce the forecast error on fast-moving time series, as data in the same cluster will have similar patterns [9]. A common method to calculate the similarity between two time series is Dynamic Time Warping (DTW) [1]. While it is an accurate measure to cluster misaligned time series that are similar in shape, it assumes that one time series is a stretched non-linear version of another time series and is expensive to compute. Moreover, previous research has only evaluated its performance and impact on the forecasting accuracy on small datasets (<1000 Stock Keeping Units (SKUs)) [18,20]. In a real world business setting, where weekly or even daily forecasts are needed, the use of DTW might be infeasible because of the computational complexity.

In this paper, we introduce a cluster-based forecasting framework that is able to forecast both intermittent and non-intermittent time series. We propose the use of Simple Exponential Smoothing (SES) as a preprocessing step before clustering. The idea behind this approach is that SES can act as a filter that removes noise from the time series and reduces outliers in the Euclidean distance, an often used distance metric in clustering. It is computationally faster to use and it can especially improve the clustering performance for intermittent time series. Smoothed time series are then less penalised by the Euclidean distance when they are misaligned, making it especially useful for intermittent time series.

We summarize our contributions as follows:

- We use a new preprocessing method before clustering to improve the clustering performance that is computationally faster to use than DTW (Sect. 3).
- We showcase our results on a publicly available dataset. Our results show that our cluster-based approach can achieve similar results in terms of the root mean squared error at a substantial lower computational cost compared to DTW (Sect. 4).
- We demonstrate that a gradient boosting machine (LightGBM) can benefit from a cluster-based approach with respect to the root mean squared error.

2   Related Work

Clustering Time Series. Cluster-based forecasting is a well studied research problem. Dividing time series into clusters can result in much smaller forecasting errors in contrast to a direct prediction [9]. The key insight in a cluster-based forecasting approach is that by partitioning the whole dataset into multiple disjoint clusters, the forecasting models trained on those separate clusters are able to improve the accuracy compared to a forecasting model trained on the whole dataset [13].

In order to cluster time series, Dynamic Time Warping (DTW) is used to as a dissimilarity metric for time series of unequal length [20]. It calculates a warping path between two sequences and is able to provide the optimal alignment. In contrast to the Euclidean distance metric, DTW can be used to cluster time series that are similar in shape but are out of phase. While DTW can be a good measure to match misaligned time series, it prove to be an infeasible method to use on a larger dataset. To calculate a distance matrix for n time series of length m, the complexity for such a operation is $\mathcal{O}(n^2m^2)$. This makes DTW an unsuitable measure in practice in terms of complexity compared to the Euclidean distance. DTW also assumes that one time series is a stretched non-linear version of another time series. In online retail, where intermittent time series are more prevalent [8], i.e. showing no clear structure, DTW is not the best suited measure to cluster intermittent time series.

Our focus is on efficiently clustering and modelling of a large dataset ($\approx 50,000$ time series) at the lowest aggregation level containing both fast-moving and intermittent time series. The proposed methods in the above mentioned works are mainly evaluated on small datasets (<1000 time series) that are forecasted at the highest aggregation level [14,20,29]. In contrast, we propose a method that foregoes the use of DTW as a distance measure but instead uses Simple Exponential Smoothing (SES) as preprocessing method to improve the clustering performance.

Time Series Forecasting. Time series forecasting has been an active area of research since the 1970s. The publication of *Time Series Analysis: Forecasting and Control* by Box and Jenkins in 1970 is perceived as an important milestone. It enabled forecast practitioners to apply a systematic approach in time series forecasting [27]. Since then, a wide diversity of methods and algorithms have been used for time series forecasting. According to [15], these methods and algorithms can be divided into two groups. The first group consists of classical statistical approaches such as SES, Auto-Regressive Integrated Moving Average (ARIMA), Theta, and Box-Jenkins. These methods are characterized as linear methods and have a strong capability of modeling trend and seasonality in a time series. The second group consists of machine learning (ML) based methods like Support Vector Regression (SVR), Long Short-Term Memory (LSTM), and tree-based models like XGBoost, LightGBM and Random Forest [15].

More recently, ML based methods are being proposed as an alternative to the classical statistical methods [2]. A good indication on the advances in forecasting theory and practice are the M competitions. The M4 competition, held in 2018, showcased that a combination of statistical models and ML models were able to outperform classical methods in terms of accuracy, highlighting the potential value of ML-based approaches in more accurate forecasting. In the M5 competition, held in 2020, exclusively ML-based methods were the winning solutions. As an example, LightGBM, a gradient-boosting framework developed by Microsoft, is able to process numerous, correlated time series effectively and reduce the forecast error [16]. In contrast to the previous M competitions, the

M5 competition was also the first competition to include intermittent demand time series.

In this paper, we use the LightGBM model as our forecasting model and conduct our experiments on the M5 competition dataset.

Forecasting Intermittent Time Series. In contrast to modelling fast-moving time series, forecasting intermittent time series is a more difficult task, as there are two sources of uncertainty: the sporadic nature of the demand volume and the sporadic timing of the demand arrivals [17]. The first proposed method to specifically forecast intermittent time series was introduced by Croston [5]. It overcomes the aforementioned difficulties by using separate series for the size of demand, and for the demand frequency. Each series is forecasted separately using Simple Exponential Smoothing (SES) and the final forecast is derived by:

$$\hat{y}_t = \frac{\hat{z}_t}{\hat{p}_t} \tag{1}$$

Where $\hat{z}_t$ and $\hat{p}_t$ are the forecast for the demand sizes and the demand intervals, respectively. Later, [24] proved that the Croston's method was biased and proposed a new method called Syntetos-Boylan Approximation (SBA): a bias-correction approximation [25]. To overcome the problem of the bias being positively correlated with the α_p smoothing parameter, a damping factor was added to Croston's method:

$$\hat{y}_t = (1 - \frac{\alpha_p}{2})\frac{\hat{z}_t}{\hat{p}_t} \tag{2}$$

While there is empirical evidence that the per series optimization can gain improvements with respect to the bias, the values for the smoothing parameter are selected in a adhoc manner in practice. Moreover, even with the recent advance in specialised methods for intermittent forecasting, simpler methods like Moving Average (MA) and SES are often used in practice [19].

In this work, we propose a clustering based approach for forecasting intermittent time series, and make use of SES as a preprocessing step.

3   Methodology

We propose a cluster-based forecasting framework that leverages Simple Exponential Smoothing (SES) as a preprocessing step. By comparing SES versions of the time series to each other, we aim to improve the clustering performance.

3.1   Framework Overview

The complete framework is illustrated in Fig. 1. Given a dataset X containing n time series of length m, we apply the SES as mentioned in Sect. 3.2 on X resulting in X'. X' is then clustered into k disjoint clusters using a clustering algorithm.

In this paper, we experiment with two clustering methods. The best performing method is selected as the clustering method in the proposed framework. Finally, a total of k LightGBM models are trained on the original non-smoothed dataset X and the predicted values are obtained.

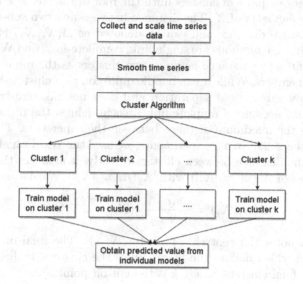

Fig. 1. The proposed framework for cluster-based time series forecasting.

3.2 Framework Details

In this section, we provide details for different components of the proposed framework. First, the smoothing step is highlighted and its effect on the clustering is discussed. We then give an overview of the two types of clustering algorithms used in our experiments. Finally, we discuss the forecasting model used to produce the predictions.

Preprocessing. As mentioned in Sect. 2, the use of DTW as a distance metric on a large dataset is computationally expensive compared to the Euclidean distance. To still be able to cluster time series that are similar in shape, exponential smoothing is used as an additional preprocessing step. The intuition behind this approach is that smoothed time series are less penalised by the Euclidean distance metric than non-smoothed time series. It enables the clustering algorithm to capture the main components of the individual time series [23]. We use SES as the smoothing approach in our framework [7]. SES is defined as follows:

$$s_t = \alpha y_t + (1 - \alpha)s_{t-1}, \tag{3}$$

where α is a smoothing factor between 0 and 1, y_t is the value of the time series at time step t and s_t is the smoothed value. The value of s_t is a simple weighted average of the current observation y_t and the previous smoothed value s_{t-1}.

Hierarchical Agglomerative Clustering. We aim to partition a dataset $X = \{x_1, ..., x_n\}$ into a collection of clusters. The first algorithm that is tried in the clustering step of the framework is Hierarchical agglomerative clustering (HAC). In HAC, records are stored at the leaves as singleton sets and the algorithm proceeds by merging pairs of clusters until the root of the tree is reached, which contains all the elements of X. The distance between any two sub-clusters of X is called the linkage distance and can be denoted as $\Delta(X_i, X_j)$ [4]. The three most common linkage methods are single-link, complete-link, and Ward's. Single-link linkage defines the distance between two clusters as the minimum distance between their members. While it can handle quite complex cluster shapes, single-link linkage only cares about separation and does not take cluster balance or compactness into account. Complete-link linkage defines the distance between two clusters as the maximum distance between their members. Therefore, it is sensitive to outliers [22]. In our experiments, we used the Ward's method. Instead of measuring the distance between clusters directly, it analyses the variance of clusters. To merge $X_i(n_i = |X_i|)$ with $X_j(n_j = |X_j|)$ Ward's method can be defined as:

$$\Delta(X_i, X_j) = \frac{n_i n_j}{n_i + n_j} \|c(X_i) - c(X_j)\|^2, \tag{4}$$

where $c(X')$ denotes the centroid of cluster X' [4]. The final output of HAC is a dendogram which defines the hierarchy of the clusters. k clusters are then selected as the final clusters, where k is the cut-off point.

Self-organizing Map. The second type of clustering algorithm used is a Self-Organizing Map (SOM). SOM is a type of artificial neural network (ANN) that is able to transform a high-dimensional input into a two-dimensional representation, called a map [11]. SOM relies on competitive learning that earns activation opportunities through competition between neurons of the output layer as opposed to error-correction learning such as back-propagation with gradient descent [3].

A SOM network consists of i neurons arranged in a 2D grid with a normally randomized weight vector m_i. The architecture allows for lateral interaction between the neurons to activate and inhibit each other. During each training iteration, a training example is fed to the network and its Euclidean distance to all weight vectors is computed. The winning neuron, also called the best-matching unit (BMU) can be expressed as:

$$W(t) = \arg\min_i \{\|x(t) - m_i(t)\|\}. \tag{5}$$

The update formula for the BMU can then be expressed as:

$$m_i(t + 1) = m_i(t) + \alpha(t) * h_{Wi}(t)[x(t) - m_i(t)], \tag{6}$$

where t is the current training iteration and x is the input vector. The amount of movement is constrained by the time-decreasing learning rate α. The learning rate is adjusted over time in order to make substantial changes in the network at the beginning phase. The neighbor neurons near $W(t)$ are denoted by the neighbor kernel h_{Wi}. In the simplest form, it is set to 1 for all neurons close enough to the BMU and 0 for others, but in practice a Gaussian neighborhood is often used [12].

Table 1. Summary statistics for the M5 dataset.

	Demand	Intermittency
Min.	0	0.0000
Median	2.0	0.1942
Max.	4220	0.8992
Mean	7.9002	0.2386
Std. Deviation	23.6665	0.1942

LightGBM. We use LightGBM as our forecasting model. Proved to be one of the best performing methods in the M5 competition [16], LightGBM is a machine learning algorithm that combines two techniques: Gradient-based One-Side Sampling (GOSS) and Exclusive Feature Bundling (EFB) in order to handle large number of data points and a large number of features.

In a gradient boosting tree, the gradient for each data instance provides important information that can be used for data sampling. If a data instance corresponds to a small gradient then the training error for this data instance will also be small, because it is already well trained. By disregarding those data instances, the distribution of the data will change, which will in turn decrease the accuracy of the model. GOSS avoids this problem by retaining all instances with large gradients and uses random sampling on those with small gradients. To counter-act the change of distribution, a constant multiplier is introduced. As high-dimensional data are usually very sparse, this sparsity of the feature space can be used to create an approach to reduce the number of features. It is possible to bundle exclusive features into a single feature because many features are mutually exclusive. This EFB step in LightGBM can significantly speed up the training of the model without decreasing the accuracy [10]. Tweedie's Poisson regression was chosen as the objective function as it deals well with right-skewed data with a large number of zeros [30], which makes is suited for the intermittent dataset.

4 Evaluation

Data. The proposed framework is evaluated on the M5 dataset from the M5 competition, ranging from January 2011 until September 2016 and aggregated at the week level. The major objective is to develop a forecasting model that can be used by the supply-chain management. Therefore, the forecasts are on the lowest aggregation level, i.e. products are forecasted at the SKU level. The dataset contains the time series for 30,472 SKUs with the length of 278. Statistical characteristics are shown in Table 1. Demand is expressed as SKU sold per week and the intermittency of a SKU is described as the fraction of zero sales, calculated from the first recorded sale in a time series. SKUs containing more than 90% of zero recorded sales are removed from the M5 dataset, containing too little information to produce any type of reliable forecast.

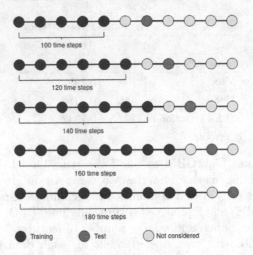

Fig. 2. Forecasting on a rolling window.

Experimental Setup. The data was split on the item level via 5-fold time series cross validation. As illustrated in Fig. 2, the forecasts are made on a rolling window basis. In each fold, the next 20 time steps are predicted. To prevent leakage of information between the training and the test set, a gap of eight time steps between the training set and test set is used. For the baseline models (Croston's Method and LightGBM) the mean, standard deviation and the median RMSE scores across the five folds are calculated.

To measure the difference in RMSE between the baseline model and the cluster forecasts on a cluster level, the predictions of the SKUs from each fold in each cluster are compared to the predictions of the SKUs in the corresponding fold of the baseline model. The mean, standard deviation, and median RMSE is then calculated to show the difference in RMSE on a cluster level. All experiments are done using an Intel Core i7-8559U CPU clocked at 2.70 GHz.

Evaluating Cluster Performance. We use internal validity measures to evaluate the performance of the clustering algorithms. The two validity measures used are the Calinski-Harabasz (CH) score and the Davies-Bouldin (DB) score. The CH score is defined as:

$$CH(C) = \frac{(N - |C|) * \sum_{c_k \in C} |c_k| d(\overline{c_k}, \overline{X})}{(|C| - 1) * \sum_{c_k \in C} \sum_{x_i \in c_k} |c_k| d(x_i, \overline{c_k})} \qquad (7)$$

With cluster scatter S denoted as:

$$S(c_k) = \frac{1}{c_k} \sum_{x_i \in c_k} d(x_i, \overline{c_k}), \tag{8}$$

the DB score of a cluster is:

$$DB(C) = \frac{1}{|C|} \sum_{c_k \in C} \max_{c_l \in C nc_k} \left\{ \frac{S(c_k + S(c_l)}{d(\overline{c_k}, \overline{c_l})} \right\}, \tag{9}$$

Where $X = \{x_1, ..., x_n\}$ is the dataset to be clustered, $C = \{c_1, ..., x_k\}$ is the clustering of X into K disjoint clusters, $\overline{c_k}$ is the centroid of the cluster and $\overline{X}$ is the centroid of the dataset [28]. The CH and DB score both measure inter and intra cluster similarities where the CH score should be maximized and the DB score should be minimized.

Evaluating Forecasting Performance. Due to the high degree of intermittency in the dataset, the Mean Absolute Error (MAE) and the Mean Absolute Percentage Error (MAPE) cannot be used. The MAE optimizes for the median and when a time series contains many zeros, i.e. intermittent, the median will also be close to zero. With intermittent time series the MAPE is also not well suited because division by zero is very likely to occur.

To evaluate the forecasting performance the Root Mean Square Error (RMSE) is used. The RMSE is defined as:

$$RMSE = \sqrt{\frac{\sum_{i=1}^{n}(y_i - \hat{y}_i)^2}{n}} \tag{10}$$

Where y_i is the actual value, $\hat{y}_i$ is the predicted value, and n is the total number of observations. Because the RMSE is scale dependent it can not be used to compare the accuracy of the models between two different datasets, however it can be used to compare the improvements in accuracy between the models on the same dataset.

4.1 Results

Time Series Clustering. Table 2 illustrates the best performing clustering configurations (number of cluster and smoothing parameter) based on the Calinski-Harabasz (CH) or Davies-Bouldin (DB) score. The parameter $\alpha = N/A$ indicates that no SES was applied before clustering. The use of SES improved the clustering results for both the HAC and SOM approaches. For the HAC method, the CH and DB score are more in agreement with what the best configuration is. A smoothing parameter of 0.1 and a cluster size of 2 or 3 both result in the highest CH score and the lowest DB score. Figures 3 and 4 show the clustering performance for the smoothing values ranging from 0.1 to 1. The CH score increases as the smoothing parameter increases, indicating that the clusters are better defined. HAC shows a maximum CH score for all ten smoothing values when 2 clusters are chosen. For the SOM the CH first increases as the the number of clusters increase with a maximum CH score between 25 and 36 number of clusters. For both clustering algorithms a smoothing parameter of 0.1 appears to result in the best defined clusters.

Table 2. Best results for different clustering methods and parameters on the M5 dataset.

Clustering method	# Clusters	α	CH score	# Clusters	α	DB score
SOM	25	N/A	1410.0694	9	N/A	3.1737
	36	0.1	2215.3234	9	0.1	1.7413
HAC	2	N/A	1163.6122	2	N/A	2.4444
	2	0.1	3229.5529	3	0.1	1.5076

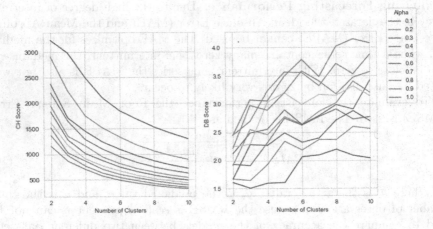

Fig. 3. Calinski-Harabasz and Davies-Bouldin scores for different smoothing parameters (alpha) using HAC.

Analyzing the cluster performance with respect to the DB score, we can see that for both types of clustering algorithms smaller clusters are preferred. The DB score is minimized for a smoothing value 0.1 for both the HAC and SOM method. For HAC the optimal number of clusters is 3 and for the SOM the optimal number of clusters is 9.

Comparison with DTW. Table 3 shows the mean and median RMSE scores for a subset of 20,000 SKUs from the M5 dataset. As a non-ML-based method, Croston's Method achieves a mean RMSE of 13.1765. A single LightGBM model is able to achieve a mean RMSE of 8.2276 across five folds and serves as the baseline model to which cluster-based models are compared. The SOM+LightGBM is

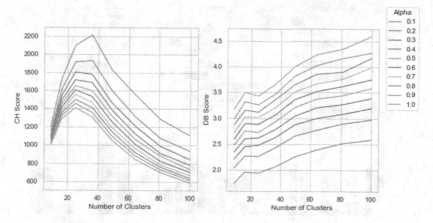

Fig. 4. Calinski-Harabasz and Davies-Bouldin scores for different smoothing parameters (alpha) using SOM.

Table 3. Mean and median RMSE scores across five folds for various models on the M5 dataset. Best performing scores are highlighted in gray.

Model	# Clusters	α	Mean	Median	CPU Time
Croston's Method	N/A	N/A	13.1765 ($\pm$ 0.9298)	12.9772	2min 15s
LightGBM	N/A	N/A	8.2276 ($\pm$ 0.3470)	8.1604	16min 13s
SOM+LightGBM	9	N/A	8.0019 ($\pm$ 0.3524)	8.0830	5min 19s
SES+SOM+LightGBM	9	0.1	8.0082 ($\pm$ 0.3371)	8.1112	6min 3s
SES+HAC+LightGBM	2	0.1	8.3030 ($\pm$ 0.3304)	8.3267	12min 57s
DTW+HAC+LightGBM	5	N/A	8.0954 ($\pm$ 0.3378)	8.0368	20h 24min 5s

the best performing model with respect to the mean RMSE, reducing the RMSE by 0,2257 (2,82%). The DTW+HAC+LightGBM model is able to achieve similar performance as the SOM+LightGBM model indicating that it could be a viable alternative. The SES+HAC+LightGBM model is not able to decrease the mean RMSE, performing worse than the baseline model. While SES is able to improve the clustering performance, it is not able to outperform the SOM model without SES with respect to the mean RMSE. Comparing CPU run times we observe that while DTW+HAC+LightGBM is able to result in similar forecasting performance, its CPU run time is multiplied by a factor of 246.

Figure 5 illustrates the mean RMSE difference per cluster as a percentage compared to the mean RMSE of the baseline model (top row). The bottom row illustrates the size of each cluster. Bars are coloured according to the mean intermittency in that cluster. Looking at Fig. 5, we observe that on a cluster level, in four out of five clusters the use of DTW+HAC were not able to produce meaningful clusters to reduce the mean RMSE. The SOM was able to reduce the mean RMSE on a cluster level for moderate (0.5) to low (0.2) average intermittency levels within a cluster. Cluster 2 and 8, having very few zero recorded sales,

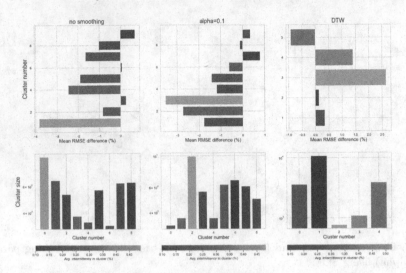

Fig. 5. Top row: mean RMSE difference across 5 folds between the baseline model (LightGBM) and the cluster-based forecasting method (SOM) on the M5 dataset. Bottom row: number of SKUs per cluster.

did not benefit from a clustered approach with respect to the mean RMSE. In contrast to DTW, the SOM was able to reduce the mean RMSE for more intermittent clusters.

Figures 6 and 7 illustrate an example forecast for an intermittent and non-intermittent time series. A clear distinction can be made between the different smoothing values. HAC is used to cluster the distance matrix made by DTW.

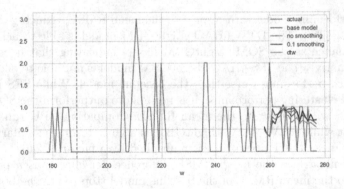

Fig. 6. Example forecast on an intermittent time series from the M5 dataset. Time series are clustered with SOM.

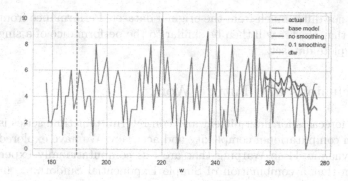

Fig. 7. Example forecast on a non-intermittent time series from the M5 dataset. Time series are clustered with SOM.

5   Discussion

Clustering time series dataset into disjoint clusters and training individual models on those clusters can lead to a reduction in the forecasting error [9,13]. A common and accurate method to measure the similarity between two time series is DTW [18], but it comes at a cost: it is a computationally expensive measure to use on large datasets and on intermittent time series, DTW may not be the best suited measure to use, since intermittent time series show no clear structure. Our results show that by applying SES as a preprocessing method, clustering and forecasting performance can be improved in combination with SOM. While not achieving the same level of performance as DTW+HAC, it does come at a substantially lower computational cost. In practice, it is then up to the forecast practitioner to decide whether the focus should lie on speed, for instance in real-time forecasting, or on accuracy when for example weekly forecasts are produced. A limitation of using SES instead of DTW is that during clustering the Euclidean distance is still used, making it not useful for time series of different length or for time series that are far apart from each other.

On a cluster level, the improvements in accuracy gained for intermittent time series seems to come at a cost of a higher forecasting error for less intermittent time series. This can be explained by the size of the produced clusters. Smaller clusters tend to perform worse than the base model because of insufficient training examples. LightGBM may also not be the best suited model to use on smaller clusters as it is sensitive to overfitting, especially on small datasets. Other models such as XGBoost or Random Forest in combination with recursive feature elimination could be used on the smaller clusters to mitigate the problem of overfitting [6].

The use of HAC, with and without SES, has almost no impact on the forecasting error, where HAC was unable to produce any meaningful clusters. HAC also suffers from *rich get richer* behaviour, which explains the uneven cluster sizes. This may also explain why on the M5 dataset, SES+HAC+LightGBM was unable to reduce the mean RMSE. A cluster containing almost all the time series does not differ much from the original dataset. Performance from a model trained on that cluster will then be similar to the performance of a single model on the complete dataset.

6   Conclusions and Future Work

Producing forecasts for intermittent and non-intermittent time series is a trade-off between computational complexity and accuracy. We have explored alternatives to Dynamic Time Warping that are less computationally expensive and have shown that a combination of Simple Exponential Smoothing and a Self-Organizing Map is able to reduce the forecast error and provide similar results to Dynamic Time Warping in combination with Hierarchical Agglomerative Clustering. Especially for intermittent time series, the use of clustering has a positive effect on the forecasting accuracy. While the use of SES as a preprocessing step was able to increase cluster performance, its effect on the forecasting accuracy was minimal. Further experimentation should clarify if other versions of Exponential Smoothing are able to decrease the forecasting error. We have also shown that LightGBM can work as a model choice when forecasts are produced for separate clusters. Future work could also explore the options of combining forecasts from a single model and a clustered model to achieve better results for both intermittent and non-intermittent time series.

References

1. Aghabozorgi, S., Seyed Shirkhorshidi, A., Ying Wah, T.: Time-series clustering a decade review. Inf. Syst. **53**, 16–38 (2015). https://doi.org/10.1016/j.is.2015.04.007

2. Ariannezhad, M., Schelter, S., de Rijke, M.: Demand forecasting in the presence of privileged information. In: Lemaire, V., Malinowski, S., Bagnall, A., Guyet, T., Tavenard, R., Ifrim, G. (eds.) AALTD 2020. LNCS (LNAI), vol. 12588, pp. 46–62. Springer, Cham (2020). https://doi.org/10.1007/978-3-030-65742-0_4

3. Chen, I.-F., Lu, C.-J.: Sales forecasting by combining clustering and machine-learning techniques for computer retailing. Neural Comput. Appl. **28**(9), 2633–2647 (2016). https://doi.org/10.1007/s00521-016-2215-x

4. Contreras, P., Murtagh, F.: Hierarchical clustering. In: Handbook of Cluster Analysis (February), 103–124. Chapman and Hall/CRC, New York (2015). https://doi.org/10.1201/b19706

5. Croston, J.D.: Forecasting and stock control for intermittent demands (1970–1977). Oper. Res. Q. **23**(3), 289–303 (1972). http://www.jstor.org/stable/3007885

6. Darst, B.F., Malecki, K.C., Engelman, C.D.: Using recursive feature elimination in random forest to account for correlated variables in high dimensional data. BMC Genet. **19**(1), 1–6 (2018)

7. Holt, C.C.: Forecasting seasonals and trends by exponentially weighted moving averages. Int. J. Forecast. **20**(1), 5–10 (2004). https://doi.org/10.1016/j.ijforecast.2003.09.015, https://www.sciencedirect.com/science/article/pii/S0169207003001134

8. Jha, A., Ray, S., Seaman, B., Dhillon, I.S.: Clustering to forecast sparse time-series data (2015)
9. Kamini, V., Vadlamani, R., Prinzie, A., Van den Poel, D.: Cash demand forecasting in ATMS by clustering and neural networks. Eur. J. Oper. Res. **232**, 383–392 (2014). https://doi.org/10.1016/j.ejor.2013.07.027
10. Ke, G., et al.: Lightgbm: a highly efficient gradient boosting decision tree. In: Advances in Neural Information Processing Systems, pp. 3146–3154 (2017)
11. Kohonen, T.: Self-organized formation of topologically correct feature maps. Biol. Cybernet. **69**, 59–69 (1982)
12. Kohonen, T.: Self-organizing feature maps. In: Self-organization and Associative Memory, pp. 119–157. Springer, Berlin (1989). https://doi.org/10.1007/978-3-642-88163-3
13. Lu, C.J., Kao, L.J.: A clustering-based sales forecasting scheme by using extreme learning machine and ensembling linkage methods with applications to computer server. Eng. Appl. Artif. Intell. **55**, 231–238 (2016). https://doi.org/10.1016/j.engappai.2016.06.015, http://dx.doi.org/10.1016/j.engappai.2016.06.015
14. Lu, C.J., Wang, Y.W.: Combining independent component analysis and growing hierarchical self-organizing maps with support vector regression in product demand forecasting. Int. J. Prod. Econ. **128**(2), 603–613 (2010). https://doi.org/10.1016/j.ijpe.2010.07.004
15. Makridakis, S., Spiliotis, E., Assimakopoulos, V.: Statistical and machine learning forecasting methods: concerns and ways forward. PLoS ONE **13**(3), e0194889 (2018)
16. Makridakis, S., Spiliotis, E., Assimakopoulos, V.: The M5 accuracy competition: results, findings and conclusions (October), pp. 1–44 (2020). https://www.researchgate.net/publication/344487258
17. Nikolopoulos, K.: We need to talk about intermittent demand forecasting. Eur. J. Oper. Res. **291**(2), 549–559 (2021). https://doi.org/10.1016/j.ejor.2019.12.046
18. Paparrizos, J., Gravano, L.: Fast and accurate time-series clustering. ACM Trans. Database Syst. **42**(2) (2017). https://doi.org/10.1145/3044711
19. Petropoulos, F., Kourentzes, N.: Forecast combinations for intermittent demand. J. Oper. Res. Soc. **66**(6), 914–924 (2015). https://doi.org/10.1057/jors.2014.62
20. Puspita, P.E., Änkaya, T., Akansel, M.: Clustering-based sales forecasting in a Forklift Distributor. UluslararasÄ Muhendislik Arastirma ve Gelistirme Dergisi, pp. 1–17, February 2019. https://doi.org/10.29137/umagd.473977
21. Seaman, B.: Considerations of a retail forecasting practitioner. Int. J. Forecast. **34**(4), 822–829 (2018). https://doi.org/10.1016/j.ijforecast.2018.03.001, https://doi.org/10.1016/j.ijforecast.2018.03.001
22. Shalizi, C.: Distances Between Clustering, Hierarchical Clustering. Data Mining (September), pp. 36–350 (2009). www.stat.cmu.edu/cshalizi/350
23. Smyl, S.: A hybrid method of exponential smoothing and recurrent neural networks for time series forecasting. Int. J. Forecast. **36**(1), 75–85 (2020). https://doi.org/10.1016/j.ijforecast.2019.03.017, https://doi.org/10.1016/j.ijforecast.2019.03.017
24. Syntetos, A.A., Boylan, J.E.: On the bias of intermittent demand estimates. Int. J. Prod. Econ. **71**(1–3), 457–466 (2001)
25. Syntetos, A.A., Boylan, J.E.: The accuracy of intermittent demand estimates. Int. J. Forecast. **21**(2), 303–314 (2005)
26. Syntetos, A.A., Zied Babai, M., Gardner, E.S.: Forecasting intermittent inventory demands: simple parametric methods vs. bootstrapping. J. Bus. Res. **68**(8), 1746–1752 (2015). https://doi.org/10.1016/j.jbusres.2015.03.034, http://dx.doi.org/10.1016/j.jbusres.2015.03.034

27. Tsay, R.S.: Time series and forecasting: Brief history and future research. J. Am. Stat. Assoc. **95**(450), 638–643 (2000). http://www.jstor.org/stable/2669408
28. Van Craenendonck, T., Blockeel, H.: Using internal validity measures to compare clustering algorithms. In: ICML, pp. 1–8 (2015)
29. Xu, S., Chan, H.K., Chng, E., Tan, K.H.: A comparison of forecasting methods for medical device demand using trend-based clustering scheme. J. Data Inf. Manag. **2**(2), 85–94 (feb 2020). https://doi.org/10.1007/s42488-020-00026-y
30. Zhou, H., Yang, Y., Qian, W.: Tweedie gradient boosting for extremely unbalanced zero-inflated data (2019)

State Discovery and Prediction from Multivariate Sensor Data

Olli-Pekka Rinta-Koski[1], Miki Sirola[1], Le Ngu Nguyen[1],
and Jaakko Hollmén[1,2(✉)]

[1] Department of Computer Science, Aalto University, Espoo, Finland
`jaakko.hollmen@aalto.fi`
[2] Department of Computer and Systems Sciences, Stockholm University,
Stockholm, Sweden

Abstract. The advent of cloud computing and autonomous data centers operating fully without human supervision has highlighted the need for fault-tolerant architectures and intelligent software tools for system parameter optimization. Demands on computational throughput have to be balanced with environmental concerns, such as energy consumption and waste heat. Using multivariate time series data collected from an experimental data center, we build a state model using clustering, then estimate the states represented by the clusters using both a hidden Markov model and a long-short term memory neural net. Knowledge of future states of the system can be used to solve tasks such as reduced energy consumption and optimized resource allocation in the data center.

1   Introduction

Cloud computing is the domain of connected and distributed computing resources, where the end user does not need to be concerned about the resource topology and nature. As systems grow in complexity, the need for intelligent, self-managing architectures becomes evident. This domain of computing systems that manage themselves autonomously according to goals set by human administrators is called autonomic computing [14]. An autonomous data center is able to operate independently, handling issues such as intermittent power failure, faulty components, and overheating without human intervention.

The AutoDC project [11] was started by a consortium of both industrial and academic partners to bring together an innovative design framework for autonomous data centers. To this end, we have investigated machine learning techniques and their applicability to problems in this domain. The machine learning goals of the project are as follows: "A powerful data analytics engine is required to achieve data collection from the various monitoring systems, which is then consolidated with external data sources and periodically stored as appropriate records to allow for both real-time and off-line ecosystem modelling and machine learning data analysis. The analytics results will ensure proper actions are applied to the control systems for optimised power, cooling, network and server operation, which is essential to maintain the data center 'health' within desired parameters to reach identified target key performance indicator (KPI) values."

© Springer Nature Switzerland AG 2021
V. Lemaire et al. (Eds.): AALTD 2021, LNAI 13114, pp. 155–169, 2021.
https://doi.org/10.1007/978-3-030-91445-5_10

In addition to the data flowing through and being processed in the data center, there is metadata being generated that has to do with the operational state of the data center itself. The amount of sensor data collected at the data center is huge, making manual annotation difficult or impossible. Our approach is to use unsupervised learning methods to organize the sensor data into states of the system, then build a model for prediction of these states. The purpose of the model is to provide an autonomous method for adjusting the control parameters of the data center, with the hope of achieving better performance in terms of resource use. Possible optimization goals include thermal efficiency and CPU utilization.

The rest of the paper is organized as follows: Sect. 2 outlines our proposed approach. Section 3 presents the clustering and prediction results from our experiments. Finally, we summarize our findings and conclude the paper in Sect. 4.

1.1 Prior Work

Previous work in this domain includes system state discovery using clustering of system log data [18] and resource usage data [7,8], novelty detection in projected spaces [25], using state change detection for history-based failure prediction [20], and forecasting for decision making support in autonomous systems [3].

Time series analysis in the data center has been applied to a number of interesting problems such as predictive maintenance, traffic balancing, and anomaly detection. Ahuja et al. [1] have used supervised learning to predict data center power variation. They have used a support vector machine [4] approach to predict power variation 15 min into the future, in order to have a reasonable time period for applying changes to control parameters. Aibin et al. [12] have used Monte Carlo tree search [16] for prediction of traffic between data centers. Shi et al. [23] have used a number of machine learning models, including long-short term memory [9], for anomaly detection from data center activity logs. Yang et al. [26] used regression methods for hardware failure prediction.

2 Methods

2.1 Summary

Our approach can be summarized as follows. First, multivariate time series data is preprocessed to reduce dimensionality. Then, system states are identified by clustering these low-dimension projections. Finally, the clustering results are used to build a model for state transition prediction. Two different approaches, hidden Markov model (HMM) [19] and long short-term memory (LSTM) [9], are discussed. Figure 1 shows a high-level view of the process for analyzing sensor data collected from a data center.

We have studied a multivariate time series data set obtained from the EDGE small data center testbed at the RISE ICE Datacenter in northern Sweden [5]. This experimental data set was created by creating a number of diverse load patterns for the purpose of cooling system performance evaluation under varying conditions. The data consists of timestamped sensor observations. Computational values include server CPU and core load. Environmental sensors include humidity and temperature.

Our model tries to identify system states based on the projection of the data onto a low-dimensional space defined by the first c principal components. Selection of c will be discussed later. These vectors are then clustered with k-means clustering [17]. These clusters are taken to represent system states. Predicting future system states makes it possible to anticipate system load and tune control parameters according to criteria such as throughput or heat emission reduction.

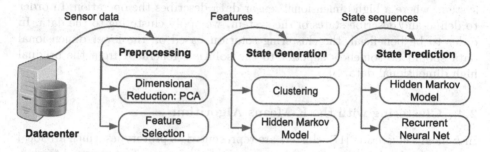

Fig. 1. A general process to analyze sensor data collected from a data center.

2.2 Data Set

Our data set contains 5072 timestamped measurements of 44 variables sampled from sensors placed in an autonomous datacenter over a time period of slightly over 42 h. The measured values include features such as power, CPU temperature, fan signal, fan power, chamber temperature, and ambient temperature. Each variable has been standardized by removing the mean and scaling to unit variance, missing values have been imputed using the mean of each variable, and the data has been resampled to uniform 30 s intervals.

2.3 Principal Component Analysis

Principal component analysis (PCA) [12] is a classical statistical technique for linear dimension reduction of multivariate data. In PCA, the goal is to represent the original d-dimensional data with a new orthogonal coordinate system that has d base vectors. By choosing a lower dimensionality starting with the most dominant base vectors, called the principal components, we can achieve a dimensionality reduction of the original data by selecting a subset of c eigenvectors. The PCA is solved by solving a corresponding eigenvalue problem. In the original data matrix $\mathbf{X} = (x_{ij}), I = 1, \ldots, n, j = 1, \ldots, d$, the entry x_{ij} denotes jth component of the ith data entry. The projection matrix $\mathbf{A}$ consisting of eigen-vectors can be used to project the original data into a new coordinate system as $y = \mathbf{A}x$.

We are interested in lowering the dimensionality of the high-dimensional sensor data in order to visualise the data in lower dimensions, and to use the lower dimensional representation as a starting point for further analysis. The phenomenon called the curse of dimensionality [24] tells us that higher dimensional data spaces are more difficult to model with finite data. Our point of investigation is to see whether the lower dimensional representation is a beneficial starting point for solving subsequent tasks.

In this paper, our goal is to extract and to describe operational states of a system, where a high dimensional sensor data describes the operation. In order to define and describe states of the system, we apply clustering to the data. In order to be beneficial, the clustering solution based on the lower dimensional data should be somehow superior to the solution generated from the original high dimensional data.

2.4   Clustering with the K-Means Algorithm

In k-means clustering [17], clusters are represented as prototypes which are local averages of the data closest to the cluster. Data points are first assigned to clusters, and the cluster memberships are iteratively adjusted according to distances to neighbouring data points.

We apply the following procedure for the analysis: We have a high-dimensional data set ($d = 44$) describing the operation of a rack in a data center and its operation environment in terms of temperature and humidity. We lower the dimensionality of the original data d to d'. Then we attempt to solve the clustering problem for k clusters ($k \in \{2, \ldots, 19\}$) using the k-means algorithm. We measure the goodness of clustering by two measures, the Davies-Bouldin index [6] and the Silhouette score [21]. The presented results are based on 5-fold cross-validation repeated 5 times.

2.5   Dynamic Modeling with Hidden Markov Model

We have used HMM [19] for estimating system state. HMM is a good fit for the problem, since the actual system state is not known but things about it can be inferred from sensor readings. For the HMM implementation, sensor outputs stand for emission observations, system states (HMM hidden states) are represented by the clusters, and transition probabilities are what we aim to estimate in order to model the dynamic behaviour of the system.

Bayesian Information Criterion (BIC) [22], which is an estimator of prediction error, was used for selecting an appropriate number of HMM states. BIC maximizes the Bayesian posterior probability.

2.6   State Prediction with Long Short-Term Memory

Recurrent neural networks are used to model the dependency of patterns in data. They have two issues: vanishing gradient and exploding gradient [10]. LSTM [9] models were introduced to resolve these issues by integrating a memory cell. The cell can capture dependencies of data over arbitrary time periods and its three gates regulate the information flow. Unlike HMMs where the Markov assumption means that the past does not affect transition probabilities, LSTMs can model arbitrary temporal dependencies.

In this work, the input of our LSTM model is a sequence of multivariate sensor data (e.g. temperature, humidity, server load, fan speed, ...). The model output is the state of the data center, which is reflected by the sensor data. Since we relied on unsupervised learning methods to analyze the data, the labels (i.e. state sequences) used to train the LSTM model were generated by the kmeans clustering algorithm and the HMM. After obtaining a sequence of states, it is possible to use a LSTM model to predict future states.

3   Results

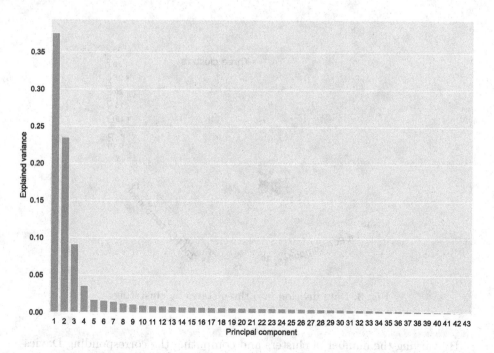

Fig. 2. Principal component scree plot.

3.1   Dimensionality Reduction with Principal Component Analysis

After using PCA to reduce dimensionality, Fig. 2 shows the explained variance of each of the principal components. It can be seen that the knee of the scree plot is at or near the fourth component. Using the Kaiser criterion [13], we retained the first four components.

3.2   Clustering with K-Means

The original data vectors of dimension 44 are replaced with vectors consisting of the first four principal components only. These are then grouped using k-means clustering. (Fig. 3 shows the data grouped into 3 clusters using the first 3 principal components.) These clusters are taken to be the states of the system. With very limited information available of the physical process, interpretation of the states is difficult and challenging. By looking at the PCA loadings in our analysis we get information about which variables are dominating in each of the PCA components, which opens opportunities to interpret the connections between states and measured variable behaviour.

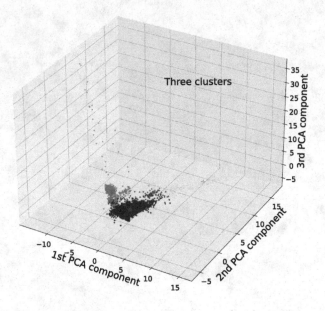

Fig. 3. Data division into three states by clustering.

By varying the number of clusters and computing the corresponding Davies-Bouldin indexes and Silhouette scores, we have found that for this particular data set, $k = 3$ achieves the best separation (Fig. 4).

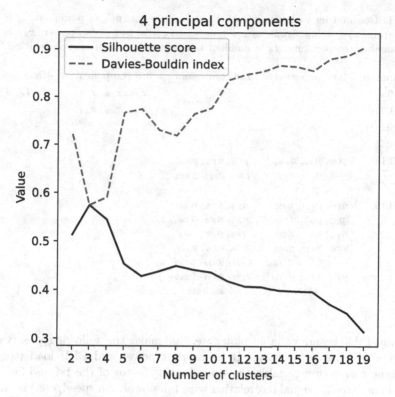

Fig. 4. Davies-Bouldin indexes and Silhouette scores for $k \in \{2, \ldots, 19\}$ clusters of vectors of 4 principal components. Lower values of Davies-Bouldin index and higher values of Silhouette score indicate better clustering.

It is possible to notice some obvious correlations between some variables and variable groups. Typically power, temperature, utilization and fan speed are associated with each other, while humidity has an inverse correlation. The effect of some variables may be delayed. Room temperature control settings may also have an effect on some variables as well as on the waste heat production of the data center.

PCA loadings tell also the variance stored in each PCA component. In our data the first four PCA components have respectively 37.1%, 23.3%, 8.7% and 2.8% of the total variance, collectively adding up to 72.0%. Table 1 shows the variables that have the largest effect on each PCA component. The scale in the left column is from 0 (no effect) to 1 (full domination). The numbers in the variables refer to the six channels. In CPU temperatures the second number refers to one of the two different cores in each channel.

Table 1. Dominating variables in each PCA component. p_{Sn} = power, y_{Smn} = CPU temperature, u_{Fn} = fan signal, n_{Fn} = number of cores, p_{Fn} = fan power, x_{Sn} = load, T_c = chamber temperature, T_a = ambient temperature, * = inverse correlation.

Dominance	1st component	2nd component	3rd component	4th component
> 0.4			x_{S5}, x_{S6}, x_{S3}	T_c^*, T_a^*
> 0.35			x_{S1}, x_{S2}, x_{S4}	
> 0.3				
> 0.2		p_{F3}, T_c^*		
> 0.18	$p_{S6}, p_{S1}, n_{F6},$ $p_{S3}, u_{F3}, n_{F2},$ p_{S4}	$p_{F1}, u_{F3}, p_{F4},$ $p_{F2}, y_{S21}^*, y_{S61}^*$		
> 0.15	$n_{F3}, y_{S21}, n_{F4},$ $p_{S2}, p_{S5}, n_{F1},$ $n_{F5}, u_{F5}, y_{S31},$ $u_{F6}, y_{S32}, y_{S61},$ $u_{F2}, y_{S11}, y_{S12},$ $u_{F4}, u_{F1}, y_{S51},$ y_{S22}	$u_{F5}, p_{F5}, n_{F4},$ $n_{F1}, n_{F2}, n_{F6},$ $n_{F3}, n_{F5}, u_{F1},$ $u_{F4}, u_{F6}, y_{S52}^*,$ $y_{S41}^*, y_{S42}^*, y_{S12}^*,$ $y_{S11}^*, y_{S31}^*, y_{S22}^*,$ y_{S62}^*, y_{S32}^*		

From Table 1 and variable plots, we can make the following observations on states and state transitions. It seems that power and CPU load (together with some correlating variables) is a dominating factor of the 1st and 2nd PCA component. The fan signal (correlating with fan power, fan speed, etc.) is another major factor in the 1st and 2nd PCA component and a reason for the smaller and bigger variations in both components. Note that the effect on the 2nd PCA component is often reversed. Variables related to utilization are strongly related to the 3rd PCA component. These variables include high peaks, see Fig. 5. Some delayed temperature and humidity changes have a small effect on the 2nd and 3rd PCA component. Temperature dominates the 4th PCA component.

The 1st and 2nd PCA components define the biggest states "idle" and "CPU power". Inside the idle state there are high peaks in the 1st and 3rd PCA components. These peaks seem to appear when we move from "idle" state to "CPU power" state. The strong vibration in fan signal especially seen in the 2nd PCA component is responsible for the third state. This effect can be noticed towards the end of the time series, see Fig. 5. Delayed temperature changes map onto different locations inside the "CPU power" state in the 3-D presentation. These effects within the state represent "cooler power on" and "hotter power on", the latter of which is more stable. The stability of CPU load also has an effect here, maybe because the average load is higher in stable load than in vibrating load. The interpretation of states affected by the 4th PCA component is made harder by the lack of graphical representation. The 4th component is mainly affected by the two temperature measurements (chamber and ambient) in an inverse relationship.

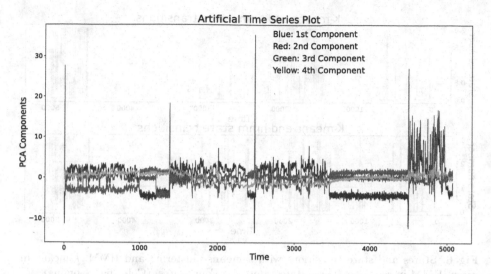

Fig. 5. PCA result in time series form.

We could label the three states as idle, high CPU load, and strongly vibrating fan signal (varying CPU load). This third state appears towards the end of the time series where the fan signal is strongly vibrating. This state is the most different to all the other states and could even be a failure state. This kind of example study demonstrates that our approach constructing states according to the physical behaviour of the process may also reveal failure states, and could be used for anomaly detection.

The described state behaviour is valid only to this type of data having a certain set of measured variables. There seem to be different state characteristics in different types of data. As an example of single measured variable types, we have experimented with large amounts of pure temperature data also collected from an experimental data center. With this data it is also possible to find similar states, but the characteristics are very different. Temperature data typically have very clear and separable clusters and states. There are also fewer characteristics such as variable delays, and the states are formed mostly by different power levels. In addition, sometimes the vibration effect forms two alternating well-separated states within one power level.

3.3 Modeling Dynamic Behaviour with HMM

HMM is used to model dynamic behaviour of the system. The state transition parameters model the changes in the system as one stable state turns into another. With HMM, the classification to a desired number of states is similar to k-means clustering. We have noted that with this data, HMM is more likely to minimize the state transitions and gives somewhat fewer transitions. The states and state transitions with both methods are seen in Fig. 6.

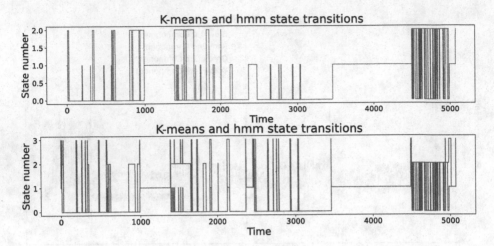

Fig. 6. States and state transitions with k-means clustering and HMM. k-means in green, HMM in red. Top: Three states. Bottom: Four states. (Color figure online)

Prediction for state probabilities for each state is calculated, see Fig. 7. Mostly one state is on (probability 1) and the other states are off (probability 0). Near the state transitions the probabilities may have also other values. One of the states only appears near the end of the time series when the fan signal is vibrating strongly.

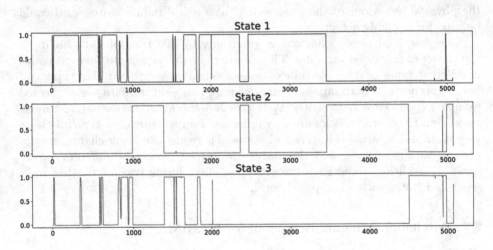

Fig. 7. State probabilities by state.

We calculated BIC according to two parameters: the number of states and the number of dimensions of the input data. We assumed that the sensor data reflected $1 \dots 20$ states of the data center and that dimensionality could be reduced while retaining enough of the relevant information. Figure 8 summarizes BIC with various numbers of states and dimensions. Our judgement was that $s \in \{3, 4, 5\}$ results in a good BIC value. This result agreed with the clustering metrics shown in Fig. 4.

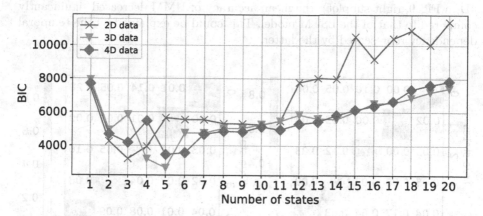

Fig. 8. Bayesian information criterion (BIC) values when varying the number of states and the number of dimensions (a model with a lower BIC is preferable).

3.4 LSTM-Based State Prediction

We performed five-fold cross-validation to obtain the aggregated results (i.e. averaging the accuracy and the confusion matrices). The LSTM model, built using the PyTorch library, had a hidden layer of dimension 20 and a time-step of 10 sequences. We train the LSTM model for 100 epochs using the Adam optimizer [15]. In each cross-validation round, the dataset was split into two parts: 80% of the samples for training and the remaining samples for testing.

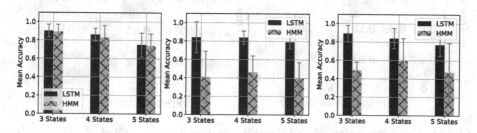

Fig. 9. Mean accuracy of hidden Markov model and long short-term memory predictions. Left: Original data of dimension 44. Middle: 3D data. Right: 4D data.

We performed the experiments with the number of states ranging from three to five. The mean accuracy of the LSTM predictions was compared to that of the

HMM trained on the same dataset (also with cross-validation). The labeling of the state sequences was generated by k-means clustering. We observed that the mean accuracy of both models was similar when trained on the original data of dimension 44 (see Fig. 9, left subplot). However, when using PCA to transform the data to a lower-dimensional space (i.e. 3D in Fig. 9, middle subplot and 4D in Fig. 9, right subplot), the mean accuracy of HMM decreased significantly compared to that of the LSTM model. This could be explained by the temporal dependency represented by the latter.

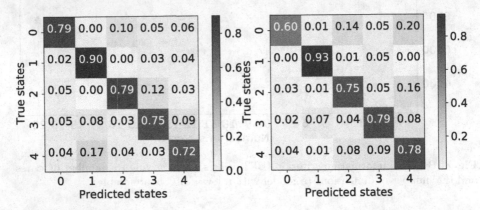

Fig. 10. Confusion matrix of LSTM predicting five states generated by clustering. Left: 3D data. Right: 4D data.

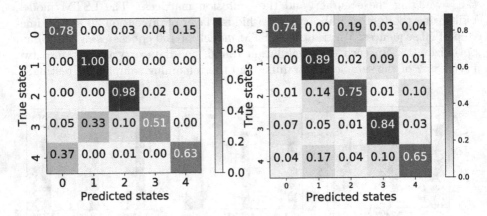

Fig. 11. Confusion matrix of LSTM predicting five states generated by HMM. Left: 3D data. Right: 4D data.

Furthermore, we investigated the predictions output by the LSTM model trained with state sequences generated by the k-means clustering algorithm (Fig. 10) and the HMM (Fig. 11). As discussed in Sect. 3.3, we decided to limit our interest to cases with the number of states $s \in \{3, 4, 5\}$. Out of these, we selected the five-state case as a representative example in which the multivariate sensor data was transformed into a 3D and 4D space using PCA. The confusion matrices showed that our LSTM model could predict the states of the data center using sequences of sensor data embedded in a lower dimensional space. We concluded that both methods (k-means and HMM) could be employed to generate labels for training state prediction models.

4   Summary and Conclusions

In this paper, we have investigated using multivariate time series data to build a model of operational states and state transitions in a data center. The data consisted of sensor output collected from sensors in the computational environment. PCA was used to reduce the dimensionality of our original data from 44 to 4, and the resulting lower-dimension data vectors were clustered using k-means to get a labeling of system states. This labeling was then used with both HMM and LSTM, building two different estimates of these states.

Our results show that unsupervised learning methods can help to discover states from multivariate sensor data. We applied k-means clustering and HMM to explore the possible states based on sensor data and achieved consistent results. We have identified a set of states of the data center, which we describe with a dynamic model. We are able to predict future states using supervised learning techniques. The use of dimension reduction techniques resulted in lowered computational complexity, which is especially useful when available resources are scarce.

Further work in this domain could involve expert interpretations of the state discovery to define how the states map onto real operational states. These results could then be applied to control data center parameters. Replicating these results with data from other data centers would make these results more generally applicable. This would require careful examination of variable selection, since the set of available sensors might be very different. Comparison with other projection techniques, such as randomized projections, remains an interesting avenue for the future.

State discovery by itself can be used to implement data visualization for monitoring. A prediction of the operational state of the data center could be used to adjust system parameters such as cooling system power or CPU allocation. Different concurrent goals such as lower power use and increased throughput may be in conflict, so a policy to optimize how the system works could be guided by intelligent algorithms that make predictions on how demands on the system will change in the short term.

Acknowledgments. We acknowledge the computational resources provided by the Aalto Science-IT project. We thank Rickard Brännvall and Jonas Gustafsson of RISE ICE Datacenter for their help with the data set.

References

1. Ahuja, N., et al.: Power variation trend prediction in modern datacenter'. In: 2017 16th IEEE Intersociety Conference on Thermal and Thermo-Mechanical Phenomena in Electronic Systems (ITherm) (2017)
2. Aibin, M., et al.: Traffic prediction for inter-data center cross-stratum optimization problems. In: Proceedings of the 2018 International Conference on Computing, Networking and Communications (ICNC): Optical and Grid Computing, p. 6 (2018)
3. Bauer, A., et al.: Time Series forecasting for self-aware systems. Proc. IEEE **108**(7) , 1068–1093, July 2020
4. Boser, B.E., Guyon, I.M., Vapnik, M.N.: A training algorithm for optimal margin classifiers. In: Proceedings of the 5th Annual ACM Workshop on Computational Learning Theory, pp. 144–152. ACM Press, Pittsburgh (1992)
5. Brännvall, R., et al.: EDGE: Microgrid Data Center with mixed energy storage. In: Proceedings of the Eleventh ACM International Conference on Future Energy Systems, E-Energy 2020: The Eleventh ACM International Conference on Future Energy Systems. Virtual Event Australia, pp. 466–473. ACM, June 12, 2020
6. Davies, D.L., Bouldin, D.W.: A cluster separation measure. In: IEEE Transactions on Pattern Analysis and Machine Intelligence, PAMI-1.2, pp. 224–227, April 1979
7. Gurumdimma, N., Jhumka, A.: Detection of recovery patterns in cluster systems using resource usage data. In: 2017 IEEE 22nd Pacific Rim International Symposium on Dependable Computing (PRDC) (2017)
8. Gurumdimma, N., et al.: CRUDE: combining resource usage data and error logs for accurate error detection in large-scale distributed systems. In: 2016 IEEE 35th Symposium on Reliable Distributed Systems (SRDS), pp. 51–60. IEEE, Budapest, September 2016
9. Hochreiter, S., Schmidhuber, J.: Long short-term memory. In: Neural Comput. **9**(8), 1735–1780 (1997)
10. Hochreiter, S., et al.: Gradient flow in recurrent nets: the difficulty of learning long-term dependencies. In: Kolen, J.F., Kremer, S.C. (eds.) A Field Guide to Dynamical Recurrent Networks, IEEE (2001)
11. ITEA3/AutoDC. About AutoDC. https://autodc.tech/about/. Accepted 8 Aug 2021
12. Jolliffe, I.T.: Principal component analysis. In: Lovric, M. (ed.) International Encyclopedia of Statistical Science. Springer, Berlin (2002). https://doi.org/10.1007/978-3-642-04898-2_455
13. Kaiser, H.F.: The application of electronic computers to factor analysis. In: Educational and Psychological Measurement, vol. 20. pp. 141–151, April 1960
14. Kephart, J.O., Chess, D.M.: The vision of autonomic computing. Computer **36**(1), 41–50 (2003)
15. Kingma, D.P., Ba, J.: Adam: a method for stochastic optimization. In: 3rd International Conference on Learning Representations (ICLR). San Diego, CA, USA (2015). arXiv: 1412.6980
16. Kocsis, L., Szepesvári, C.: Bandit based Monte-Carlo planning. In: Fürnkranz, J., Scheffer, T., Spiliopoulou, M. (eds.) ECML 2006. LNCS (LNAI), vol. 4212, pp. 282–293. Springer, Heidelberg (2006). https://doi.org/10.1007/11871842_29
17. Lloyd, S.: Least squares quantization in PCM. IEEE Trans. Inf. Theory **28**(2), 129–137 (1982)

18. Makanju, A., Zincir-Heywood, A.N., Milios, E.E.: System state discovery via information content clustering of system logs. In: 2011 Sixth International Conference on Availability, Reliability and Security (ARES), pp. 301–306. IEEE, Vienna, August 2011
19. Rabiner, L.R.: A tutorial on hidden markov models and selected applications in speech recognition. Proc. IEEE **77**(2), 257–286 (1989)
20. Rajachandrasekar, R., Besseron, X., Panda, D.K.: Monitoring and predicting hardware failures in HPC clusters with FTBIPMI. In: 2012 IEEE 26th International Parallel and Distributed Processing Symposium Workshops & PhD Forum, pp. 1136–1143. IEEE, Shanghai, China, May 2012
21. Rousseeuw, P.J.: Silhouettes: a graphical aid to the interpretation and validation of cluster analysis. J. Comput. Appl. Math. **20**, 53–65 (1987)
22. Schwarz, G.: Estimating the dimension of a model. Ann. Stat. **6**(2), 461–464 (1978)
23. Shi, J., He, G., Liu, X.: Anomaly detection for key performance indicators through machine learning. In: 2018 International Conference on Network Infrastructure and Digital Content (IC-NIDC), pp. 1–5. IEEE, Guiyang, Aug. 2018
24. Steinbach, M., Ertöz, L., Kumar, V.: The challenges of clustering high dimensional data. In: Wille, L.T. (ed.) New Directions in Statistical Physics, pp. 273–309. Springer, Berlin, Heidelberg (2004). https://doi.org/10.1007/978-3-662-08968-2_16
25. Toivola, J., Prada, M.A., Hollmén, J.: Novelty detection in projected spaces for structural health monitoring. In: Cohen, P.R., Adams, N.M., Berthold, M.R. (eds.) IDA 2010. LNCS, vol. 6065, pp. 208–219. Springer, Heidelberg (2010). https://doi.org/10.1007/978-3-642-13062-5_20
26. Yang, W., et al.: Hard drive failure prediction using big data. In: 2015 IEEE 34th Symposium on Reliable Distributed Systems Workshop (SRDSW), pp. 13–18. IEEE, Montreal, QC, September 2015

RevDet: Robust and Memory Efficient Event Detection and Tracking in Large News Feeds

Abdul Hameed Azeemi[1]([✉]), Muhammad Hamza Sohail[1], Talha Zubair[1], Muaz Maqbool[1], Irfan Younas[1], and Omair Shafiq[2]

[1] FAST-NUCES, Lahore, Pakistan
{1154031,1154074,1154166,1154053}@lhr.nu.edu.pk, irfan.younas@nu.edu.pk
[2] Carleton University, Ottawa, Canada
omair.shafiq@carleton.ca

Abstract. With the ever-growing volume of online news feeds, event-based organization of news articles has many practical applications including better information navigation and the ability to view and analyze events as they develop. Automatically tracking the evolution of events in large news corpora still remains a challenging task, and the existing techniques for Event Detection and Tracking do not place a particular focus on tracking events in very large and constantly updating news feeds. Here, we propose a new method for robust and efficient event detection and tracking, which we call RevDet algorithm. RevDet adopts an iterative clustering approach for tracking events. Even though many events continue to develop for many days or even months, RevDet is able to detect and track those events while utilizing only a constant amount of space on main memory. We also devise a redundancy removal strategy which effectively eliminates duplicate news articles and substantially reduces the size of data. We construct a large, comprehensive new ground truth dataset specifically for event detection and tracking approaches by augmenting two existing datasets: w2e and GDELT. We implement RevDet algorithm and evaluate its performance on the ground truth event chains. We discover that our algorithm is able to accurately recover event chains in the ground-truth dataset. We also compare the memory efficiency of our algorithm with the standard single pass clustering approach, and demonstrate the appropriateness of our algorithm for event detection and tracking task in large news feeds.

Keywords: Event detection and tracking · Large news feeds · Event chains

1 Introduction

Internet today has become the primary source for creation and widespread dissemination of news articles leading to generation of huge amounts of news data each day. With this unprecedented increase in the information available online,

V. Lemaire et al. (Eds.): AALTD 2021, LNAI 13114, pp. 170–185, 2021.
https://doi.org/10.1007/978-3-030-91445-5_11

one of the major challenges is providing the user with better information navigation capability. In this scenario, automatic event-based organization of news data can lead to better structuring and classification of textual news articles data from a variety of online news media sources, and thus provide users with a better online experience. The task of automatic, event-based organization of textual news article data is named as event detection and tracking (also referred to as topic detection and tracking in many contexts) [1,2]. The process of discovering a new event in a stream of news articles is referred to as Event Detection. Event tracking involves the identification of further news stories that discuss the detected event, and provide some additional information indicating that the event has developed. Hence, the major task of event tracking techniques is in essence the identification of relationships between the news articles based on the event they report [2,6].

The existing techniques for Event Detection and Tracking do not place a particular focus on tracking events in very large, complex and constantly updating news feeds.

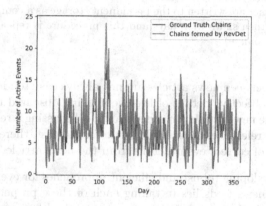

Fig. 1. Per day active event chains of an year formed by our RevDet algorithm vs the ground truth chains. To form these chains, RevDet only utilized memory required for storing eight days data.

The major challenge of applying Event Detection and Tracking techniques to very large news feeds is coping with the Variety, Velocity and Volume (3V's of Big Data) of such databases. Large and constantly updating news feeds exhibit the following properties:

1. Most of the events occurring across the globe are reported by multiple news agencies adding a great deal of redundancy in news feeds and a significant increase in volume.
2. News articles reporting a rapidly developing event tend to occur in bursts and are similar in the mention of locations *i.e.* they exhibit strong spatio-temporal correlation.

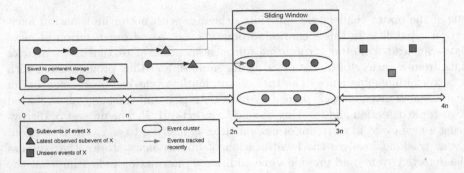

Fig. 2. RevDet maintains a sliding window of size n for performing memory efficient event tracking. For a sample event X, at any given time, only the latest observed subevents (of the event X) and the events in the sliding window are kept in the main memory. The events inside the window are clustered together through the Birch clustering method. After clustering, similarity between the earliest events in these clusters and the latest representatives (represented by triangles) is computed. Similar sub-events are joined together (represented by red arrow) to form an event chain. The events which are not tracked further are written to the permanent storage as a complete event chain. The event window then slides by n days and this procedure is repeated until the last event. (Color figure online)

3. Relationships between news articles are not always easy to identify from their text, with the objective details of the event being obscured by the reporting style used in the news article *e.g.* a news article discussing a recently occurring event may give references to multiple events in the past, thereby complicating the extraction of correct event details from the news article.

The key to developing robust and efficient approaches for event detection and tracking in large news feeds lies in taking each of these properties into special consideration. In this paper, we propose a new method for event detection and tracking, which we call the RevDet algorithm. In our method, we adopt an iterative clustering approach for tracking events by using only a constant amount of space (Fig. 2). Even though many events continue to develop for many days or even months, our method is able to track such events and form chains with a window-size set to a small time unit of eight days. We also devise a redundancy removal strategy which effectively eliminates duplicate news articles and substantially reduces the size of data. Moreover, instead of utilizing all of the content of news articles, we develop a concise representation using only the article's title and a list of locations. For evaluating our algorithm, we also construct a large, comprehensive new ground truth dataset by augmenting two existing datasets: w2e and GDELT. We implement RevDet algorithm, and evaluate its performance on the ground truth event chains. We discover that our algorithm is able to accurately recover event chains in the ground-truth dataset (Fig. 1), with precision of 0.82 and an F_1 score of 0.66. We also compare the memory efficiency of our algorithm with the standard single pass clustering approach and demonstrate the appropriateness of our algorithm for event detection and tracking task in large news feeds.

2   Related Work

The task of Topic Detection and Tracking (with the topic meaning an event) was first conceived in 1996 and evolved as a joint venture between University of Massachusetts, Carnegie Mellon University and Dragon Systems [1]. It was a yearlong pilot study focusing on segmentation of data streams, identifying events in news stream and tracking a particular event in different news. This initiative provided grounds for further research on this topic and established some initial techniques and methodologies to address the problem.

The problem was divided into three main tasks:

1. Segmentation of the data/news stream into distinct, topically homogenous blocks.
2. Identification the first occurrence of a news story discussing a new event.
3. Subsequent tracking of the news stories that discuss the event.

Existing event detection and tracking algorithms usually adapt the single pass clustering algorithm for the identification of news events [3,6,11]. For each incoming news article, its similarity with previous known events is computed. If the similarity exceeds a similarity threshold, the news article is flagged as referring to an existing event. Otherwise, the news article is classified to be a new event. The inherent problem in the application of single pass clustering algorithm to very large event data is evident: the single pass clustering algorithm must maintain a "memory" of news events. Although this is feasible for small datasets such as TREC, maintaining all the events in memory quickly becomes a significant challenge if large news feeds are dealt with, due to the scale at which events are reported each day all over the world. Alternate forms of news representation such as forming a query with only named entities and quantitative details fails to address the problem; important details form a significant part of news articles, and rigorous preprocessing for a significant reduction in size of news article in memory can lead to misclassification and a significant drop in precision.

The solution to this problem is to extend the concept of a growing 'entropy' of the news article as used by Radinsky [13] i.e. penalizing an event on the time distance between two events. Experiments on the TDT4 corpus with different time thresholds have shown $n = 14$ days threshold to be the most appropriate. If we make this a binary threshold, we will need to place only n days data in the memory, with n being the upper limit on the number of days between any two events as determined by experiments.

Other systems have considered tracking more generic 'topics' within the news articles. In [10] a framework has been presented for tracking topics in news articles via short, distinct phrases that remain intact throughout the articles. Our focus, however, is to develop a technique for tracking 'events' instead of 'topics' in news stories, which are more specific and require a much greater context than just a few phrases for achieving a high precision. Some efforts have been made to leverage topic modelling for detection and tracking of news events. One such technique, Latent Dirichlet allocation (LDA), is widely used in detecting events through posts on micro-blogging sites such as Twitter. Diao et al. [4] developed an LDA model which is able to find bursty topics on Twitter by capturing two phenomena: posts by same user or around same times are more likely to correspond to same topic/event.

Graph-based modeling approaches have also been used in Event Detection and Tracking. Sayyadi et al. [14] presented an approach of building a KeyGraph with keywords having lower inverse document frequency (IDF) filtered out. Connected key words are those that occur in same document. Closely related words form a community. A community is considered to be a synthetic document and titled as a key document. Clustering then groups together documents similar to the key documents, and each cluster is considered as an event.

Many event detection and tracking algorithms tend to perform better on carefully curated test datasets but struggle to generalise to real world news feeds. To the best of our knowledge, this is the first attempt to devise an event detection and tracking strategy for large, noisy and complex news feeds containing a great portion of duplicate news articles (Fig. 3).

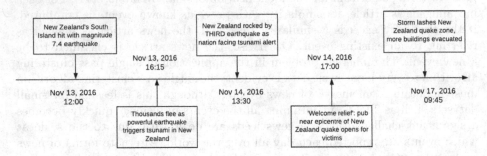

Fig. 3. An event chain showing the progress of subevents related to earthquake in New Zealand.

3   Definitions

Event. An event is an occurrence at a particular location during a particular interval of time. An event is further composed of **subevents** such that the beginning and the end of an event correspond to two separate subevents. Since we are dealing with online news data, we will consider newsworthy events only *i.e.* events that are significant enough to be reported by at least one online news agency.

Subevent. A subevent, which is an atomic part of an event, is an occurrence at a particular location and time. A subevent may only be a part of one event only.

News Article. A news article a represents a subevent e is characterized by its publication timestamp t, title h and a list of locations mentioned in the article l.

$$a = (t, h, l) \tag{1}$$

Event Chain. An event chain C is an ordered set of sub events $\{e_1, e_2, e_3, ..., e_n\}$ of a particular event, sorted in increasing order of timestamp and where each new sub event has some additional information as compared to its predecessor.

$$C = \{a_1, a_2, a_3, ..., a_n\} \tag{2}$$

Latest Representative (LR). Latest representative of an event a_1 is the sub-event in event chain with the latest timestamp *i.e.* the most recent news about an event.

Earliest Representative (LR). Earliest representative of an event a_n is the sub-event in event chain with the smallest timestamp *i.e.* the first news about an event.

Event Window. Event window consists of unordered subevents of different events occurring in a particular time frame Δt.

4   Approach

The first step in devising an approach for event tracking is to consider what makes an event different from others. Depending on this definition of an event, the event chains formed may be considerably different *e.g.* an event chain of a general election in a certain country may involve all the news in relation to the election or only the news relating to the rallies by one candidate. The decision of this is made by determining what constitutes the event identity [2], which is something unique to every new event, and common to the sub-events in event chain. If an event is taken to be something that happens at particular place and time, then the locations mentioned in a news story and $t \pm n$ days constitutes the identity of event, with t being the event timestamp. Another option could be to include named entities *e.g. people, organizations* as part of event identity, and this has been seen to considerably increase recall in event tracking tasks [13].

Selecting the Clustering Algorithm. The choice of clustering algorithm for the formation of event chains is an important one, since it directly influences the representation of news articles, quality of chains formed and efficiency of the approach. Some approaches have used the k nearest neighbours algorithm for finding closest news articles or the k-means algorithm for grouping together the related news. These methods would fail to work in a big data setting since they require a parameter k as input *i.e.* the number of articles to group together. Other algorithms include Wave-Cluster, DBSCAN and BIRCH. Wave-Cluster is a grid based algorithm and the main advantage of this algorithm is the fast processing time [5]. However, Wave-Cluster does not perform well for our problem as using a single uniform grid does not result in good quality clusters nor does it satisfy the time constraints for a highly irregular data distribution (news articles). DBSCAN is a density based clustering algorithm. It can efficiently deal with noise while forming high quality clusters. Unlike k-means, DBSCAN does not require the input parameter k which is used to identify the number of clusters to be formed. Although, DBSCAN seems to be a good choice, the major drawback of this algorithm is its inability to efficiently cluster data sets with large differences in densities.

BIRCH [15] is an unsupervised hierarchical clustering algorithm suitable to cluster large data sets. The main advantage of using BIRCH is that it can work incrementally *i.e.* does not require the whole data set in advance and can efficiently adjust the number of clusters to be formed relative to the input data set. BIRCH typically requires a single scan of the data set to form good quality clusters and this quality can be improved using additional scans if required.

Similar to DBSCAN, BIRCH can work without the input parameter k and can decide for itself the number of clusters to be formed. This feature of BIRCH is essential for our research problem as the number of event chains present in a given set of news articles can vary. Moreover, BIRCH is a first of its kind algorithm that can efficiently handle noise. News articles which do not progress are considered noise in our case as they form an event chain consisting of only one node i.e. they are not tracked further.

Representation of News Articles. We represent every news article as a vector of title, themes, locations and counts contained within the news article.

1. **Title** of the news article reporting the event.
   Example: Powerful earthquake strikes New Zealand killing 2 people.
2. **Themes** associated with the event.
   Example: NATURAL DISASTER; NATURAL DISASTER EARTHQUAKE; CAUTION ADVICE; KILL;
3. **Locations** contained within the news article.
   Example: Wellington, New Zealand, (Lat, Lng): -41.3,174.783
4. **Counts** associated with the event reported by the article, and of a particular location.
   Example: KILL 2, New Zealand, NZ;

These fields have been pre-extracted for every article in GDELT GKG. Instead of a *tf-idf* representation, we convert themes, locations and counts into one-hot vectors, and use a sparse representation of these vectors. This type of representation is readily accepted by the existing implementations of the Birch algorithm *e.g.* SciKit [12] implementation of Birch.

Figure 4 gives an overview of the workflow adopted for forming event chains on the prepared dataset through the proposed algorithm, and evaluating the results.

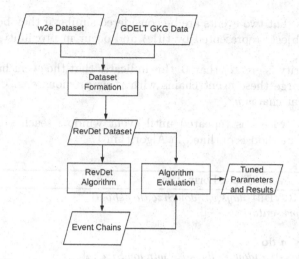

Fig. 4. A high level overview of the approach taken for formation of event chains and evaluation of results

5 RevDet Algorithm

Now, we describe our algorithm to form event chains from news data. We say that every sub event x in an event chain C_i contains sufficient information that enables tracking of further events solely through x, and that these subevents cannot track events of some other event chain C_j. i.e. for every a, b in an event chain C_i,

$$similarity(C_a^i, C_b^i) > \theta > similarity(C_a^i, C_x^j) \mid i \neq j$$

In other words, if we are presented only with the first event of a chain, we will be able to recover the whole event chain from the news feed. We adopt an iterative clustering approach for tracking events.

1. Initially we add first n days data to the event window.
2. Then we cluster articles data through the birch clustering algorithm, and save the resultant event chains to permanent storage.
3. Now, we extract the latest representatives of these event chains, and keep them in temporary storage, discarding the rest of data in the chains (at any given time, we only keep latest representatives belonging to at most one event window in the past). We slide the event window by n days.
4. We then again cluster articles data to form chains y. For each latest representative l_i saved in the previous step, we compute its similarity with each of the earliest representative e of y.

$$sim(l_i, e)$$

where the similarity of two events $sim(a, b)$ is defined as:

$$jaccard(a_{title}, b_{title}) * jaccard(a_{location}, b_{location})$$

This ensures that two events are be considered similar if they both belong to a certain subject (represented by title) and occur in proximity (represented by location).

5. If the similarity is greater than 0, this indicates that the event has developed; hence we merge these event chains with their previous one. Otherwise, we save the event chains y.

This whole process is repeated until event window reaches the end. The overall RevDet method is outlined in Algorithm 1.

Algorithm 1. Event Chain Formation

1: **procedure** REVDET(*days,windowSize,threshold*)
2:     $latestRepresentatives \leftarrow []$
3:     $i \leftarrow 0$
4:     **while** $i \leq n$ **do**
5:         $previousWindow \leftarrow days[i - windowSize : i]$
6:         $latestRepresentatives.keep(previousWindow)$
7:         $end \leftarrow i + windowSize$
8:         $data \leftarrow getData(days[i : end])$
9:         $df \leftarrow concat(data['title'], data['locations'])$
10:         $df \leftarrow oneHotEncode(df, sparseOutput = true)$
11:         $clusters \leftarrow birchClustering(df, threshold)$
12:         **for** cluster in clusters **do**
13:             $eR \leftarrow getEarliestRepresentative(cluster)$
14:             **for** row in latestRepresentatives **do**
15:                 $s1 \leftarrow jaccardSimilarity(row.title, eR.title)$
16:                 $s2 \leftarrow jaccardSimilarity(row.location, eR.location)$
17:                 **if** $s1 > 0$ and $s2 > 0$ **then**
18:                     $connectedEvents \leftarrow getEventChainByID(eR.id)$
19:                     $df \leftarrow concat(connectedEvents, df)$
20:                     $latestRepresentatives.remove(row)$
21:             $lR \leftarrow df.tail()$
22:             $latestRepresentatives.concat(lR)$
23:             $df.sort()$
24:             $saveEventChain(id = lR.id, data = df)$
25:         $i \leftarrow i + windowSize$

5.1 Implementation

We have implemented the RevDet algorithm in Python on top of the Birch Clustering Algorithm available in SciKit Learn [12]. Our algorithm takes as input news articles data (with two necessary columns: a list of locations and title) in the form of per day files (sorted by ascending timestamp of the event), window size and birch threshold. It then forms event chains and outputs each chain in a separate file. During the formation process, it also writes some temporary files to the permanent storage, and removes them once all chains have been formed.

6 Experiments

6.1 Dataset

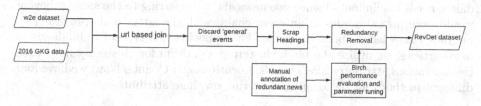

Fig. 5. Formation of the RevDet dataset for evaluating event detection and tracking approaches.

GDELT

GDELT [9] is a real-time database of global human society, and essentially contains a large amount of processed world news . The GDELT global knowledge graph (GKG) is a part of GDELT database, and is the largest publicly available dataset of news events across the globe. It contains processed data from real-time news from around the world including locations, themes, organizations, people and tone of every news event. The GKG table in the GDELT database has 27 columns containing a wealth of information about each news article. This dataset provides us with pre-extracted fields of each news article for running our Event Detection and Tracking algorithm.

Along with this, we require a fairly large event tracking dataset with fine-grained ground truth for an effective evaluation of our algorithm. TREC's TDT's datasets are unsuitable for this purpose, as they are obsolete and small: they were collected in the year 2000 and have around only 13k articles grouped in 279 topics. The recently released dataset w2e [7] is a manually constructed substantially large TDT dataset containing 207,722 events grouped in 4501 events and 2015 event chains. Each event chain contains urls of news articles and short text describing each subevent in the chain.

Although w2e contains a short description of each event, it lacks the specific processed details of news events as available in GKG (themes, locations, tone etc.). To address this problem, we reconstruct the w2e dataset by augmenting it with the GDELT dataset *i.e.* each url in the original w2e dataset is searched in GDELT GKG table, and the details contained in the matched row in GKG table are appended to w2e. From the resultant data, we keep only the chains which adhere to the concept of event defined earlier *i.e.* throughout its development, a news event must contain similar locations. This process discards chains with a more general topic for example a chain containing all news related to the US Presidential Election, instead of a specific event. Following this process (Fig. 5), we are able to construct a fairly large and a rich dataset: RevDet dataset, for evaluation of our event tracking algorithm containing 1329 event chains.

6.2 Redundancy Removal

Most of the events are cited by multiple news agencies across the globe, thereby adding a substantial amount of redundancy to data in news feeds. This redundancy needs to eliminated since two news articles referring to the same subevent would occur as two nodes in an event chain, with the latter node providing no upgraded knowledge about the event. For removing this type of redundancy in news articles, we utilize the birch clustering algorithm for clustering news articles on various attributes like Themes, Locations and Counts. Now, we have four different methods for performing clustering on these attributes:

1. Clustering on title and locations first, then sub-clustering the resulting clusters on the basis of counts,
2. Clustering on title, then sub-clustering on locations and counts,
3. Clustering on title, locations and counts, or
4. Clustering on locations, then sub-clustering on title and counts

To compare the performance of these four methods and tune birch parameters, we manually cluster a subset of GKG data of 354 news articles containing 7 events and construct a ground-truth dataset containing clusters of duplicate news articles. Two news are grouped together only if they represent the exact same subevent. It is important to note here that while they contain the same information, they are two different news articles with possibly different reporting styles and the choice of words. Hence, our task is tailored towards news data and slightly different from the approaches for near-duplicate detection, which are more general and do not consider specific properties of news articles like title, locations and counts etc.

We evaluate performance by clustering the news articles and comparing to the ground truth clusters. Clustering accuracy is evaluated by calculating Precision, Recall and F_1-Score over pairs of articles $i.e.$ through the pair-counting method. The precision is calculated as

$$P = \frac{TP}{TP + FP}$$

$i.e.$ the fraction of pairs correctly put in one cluster, and recall as

$$R = \frac{TP}{TP + FN}$$

$i.e.$ how many actual pairs were identified. F_1-score is the harmonic mean of precision and recall and is used for selecting the best birch parameters for each clustering approach, and we use this score for comparing the four clustering approaches (Table 1). As shown, clustering on title and locations first, then sub-clustering on counts yields the best result making it a suitable approach for removing duplicate news articles. This procedure results in a 57% decrease in the data size.

Table 1. Precision, recall and F_1 score for four different approaches of clustering redundant news. Clustering on title and locations first, and then sub-clustering on counts yields the best result, implying that the title and locations combined have the greatest discriminatory power of correctly separating two different news.

First level	Second level	Precision	Recall	F_1
Title, Locations	Counts	0.97	0.77	0.86
Title	Locations, Counts	0.75	0.79	0.77
Title, Locations, Counts	–	0.67	0.68	0.67
Locations	Title, Counts	0.92	0.41	0.57

Article's Title vs Content. We now consider using article's content instead of title for detecting duplicate news articles to see whether there is a significant gain in the F_1-score. For this task, we use the themes field (originally contained in GDELT GKG) in the dataset, which describes all the themes contained in a news article through special categories and taxonomies which accurately capture the content *e.g.* a news article about the destruction of roads by heavy rain contains themes like

- NATURAL_DISASTER_MONSOON
- INFRASTRUCTURE_BAD_ROADS

We compare the performance by first clustering duplicate news on title, locations and then counts. We repeat the same process with themes instead of title. As the results in Table 2 show, using article's content (themes) does not lead to a significant change in the F_1 score. This shows that a news article's title has the ability to accurately and succinctly describe the event reported in it. Moreover, as the average content length of a news article in the data (represented by themes length) is significantly greater than the article's title, clustering on title is a more suitable option of removing duplicate news than clustering on article's content.

Table 2. Comparing the performance of title and content (represented by themes) for clustering duplicate news together.

Method	Av. length in characters	Precision	Recall	F_1 score
Title	18.7	0.97	0.77	0.86
Themes	16476.0	0.96	0.81	0.88

6.3   Algorithm Evaluation

We evaluate the performance of the algorithm by comparing the event chains in the ground-truth dataset with the event chains formed by the algorithm. For this, we first transform the dataset into per day files, simulating the way in which data would be available to the algorithm in a news feed (Fig. 6). We then

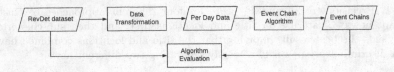

Fig. 6. An overview of the steps involved in preparing data for evaluation of event chains formed by RevDet

run RevDet on these per day files and evaluate performance through Precision, Recall and F_1 score over pairs of articles in the ground truth event clusters and the formed clusters.

Clustering Performance. The best performance of RevDet is reached on Birch Threshold 2.3, and Window Size (Table 3). The F_1 score of 0.66 on 0.82 precision is adequate enough to form event chains of good quality as have focused on precision focused tuning to avoid distortion of event chains with irrelevant news. A relatively low recall indicates the difficulty in clustering news events together with different wordings of the title. This problem can be alleviated in future work through the use of pretrained paragraph level embeddings like Doc2Vec [8]. The in-memory clustering approach has a much lower precision and recall. This is due to a greater chance of an event landing in the wrong chain as the time-dependancies between events are ignored by loading all the data into the memory at start and then performing clustering.

We next present a macro-level comparison of active event chains in the original dataset and in the formed ones in Fig. 1, which shows the number of event chains that are still being developed on each day. It can be seen that RevDet has been able to closely replicate the ground truth events.

Table 3. RevDet vs In-Memory clustering performance on tuned parameters as evaluated on ground truth chains. RevDet performs far better than the in-memory clustering approach.

Algorithm	Birch threshold	Window size	Precision	Recall	F_1
RevDet	2.3	8	0.81	0.56	0.66
In-Memory	2.2	–	0.56	0.24	0.34

Window Size. We next focus on the effect of window size on the results (Fig. 7). We discover that varying the window size after 8 has little effect on the F_1 score *i.e.* it stays between 0.64 and 0.66. This makes 8 a good choice for window size, and implies that most of event chains do not have a gap of greater than 8 days between any two consecutive news. We also observe that the precision drops slightly as the window size is increased, owing to the greater data in the event window.

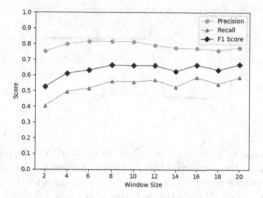

Fig. 7. Plot of precision, recall and F_1 score vs. window size of RevDet. At window size 8, RevDet is able to track events with almost same clustering accuracy as with window sizes closer to 20, while needing much lesser memory.

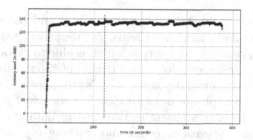

Fig. 8. Memory usage vs running time of RevDet algorithm. The small spikes represent the movement of event chains to and from the main memory according to their development.

6.4   Scalability of RevDet

We next examine the memory efficiency and scalability of RevDet. The plot in Fig. 8 shows the memory usage as the algorithm progresses. As expected, the space requirement of temporary storage (RAM) is constant with respect to the input data. The spikes are representative of the movement of event chains to and from the memory. RevDet has the ability to scale efficiently with respect to the number of news articles in the dataset which makes it a very suitable approach for event detection and tracking in large news feeds. We also examine the space requirement of the in-memory clustering approach in Fig. 9. The memory usage rises sharply as all of news data is loaded into the main memory at the start, and becomes constant once formed chains are being written to the permanent storage. Peak memory usage of the in-memory clustering approach ($\approx$1000 MB) is 7 times the peak memory usage of RevDet ($\approx$140 MB). Moreover, as the input data will increase, the memory requirements of the former approach will grow proportionally making it infeasible to form event chains.

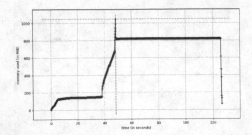

Fig. 9. Memory usage vs running time of an in-memory clustering approach which loads all the data into the memory once and then performs clustering. The memory increase from 40 s to 50 s represents the transfer of all data to the memory; the spike at 50 s is due to clustering all data through Birch at once.

7   Conclusion and Future Work

In this paper, we have tackled the problem of robust and efficient detection and tracking of news events in large news feeds. An iterative clustering based algorithm has been proposed for this purpose which is able to extract event chains of events that continue to develop for a long period of time, using memory as low as required for clustering eight day news. We also propose a redundancy removal strategy for removing duplicate news articles. We construct a new, comprehensive ground truth dataset by augmenting two existing datasets: GDELT and w2e, specifically for evaluating event detection and tracking approaches. We show the efficacy of our method by evaluating it on the ground-truth chains. We leave for future work the improvement in recall by clustering news articles through incorporation of more robust text representations like Doc2Vec. RevDet can also be extended easily to work for streaming news data and this can lead to a truly automated and robust event classifier and an event search engine.

References

1. Allan, J., Carbonell, J.G., Doddington, G., Yamron, J., Yang, Y.: Topic detection and tracking pilot study final report (2003)
2. Allan, J., Papka, R., Lavrenko, V.: On-line new event detection and tracking. In: SIGIR, vol. 98, pp. 37–45. Citeseer (1998)
3. Becker, H., Naaman, M., Gravano, L.: Beyond trending topics: real-world event identification on Twitter. In: Fifth International AAAI Conference on Weblogs and Social Media (2011)
4. Diao, Q., Jiang, J., Zhu, F., Lim, E.P.: Finding bursty topics from microblogs. In: Proceedings of the 50th Annual Meeting of the Association for Computational Linguistics: Long Papers, vol. 1, pp. 536–544. Association for Computational Linguistics (2012)
5. Fahad, A., et al.: A survey of clustering algorithms for big data: taxonomy and empirical analysis. IEEE Trans. Emerg. Top. Comput. **2**(3), 267–279 (2014). https://doi.org/10.1109/TETC.2014.2330519

6. Hasan, M., Orgun, M.A., Schwitter, R.: Real-time event detection from the twitter data stream using the TwitterNews+ framework. Inf. Process. Manage. **56**(3), 1146–1165 (2019)
7. Hoang, T.A., Vo, K.D., Nejdl, W.: W2E: a worldwide-event benchmark dataset for topic detection and tracking. In: Proceedings of the 27th ACM International Conference on Information and Knowledge Management, pp. 1847–1850. ACM (2018)
8. Le, Q., Mikolov, T.: Distributed representations of sentences and documents. In: Proceedings of the 31st International Conference on International Conference on Machine Learning, ICML 2014, vol. 32, pp. II-1188–II-1196. JMLR.org (2014). http://dl.acm.org/citation.cfm?id=3044805.3045025
9. Leetaru, K., Schrodt, P.A.: GDELT: global data on events, location, and tone, 1979–2012. In: ISA Annual Convention, vol. 2, pp. 1–49. Citeseer (2013)
10. Leskovec, J., Backstrom, L., Kleinberg, J.: Meme-tracking and the dynamics of the news cycle. In: Proceedings of the 15th ACM SIGKDD International Conference on Knowledge Discovery and Data Mining, pp. 497–506 (2009)
11. Osborne, M., et al.: Real-time detection, tracking, and monitoring of automatically discovered events in social media (2014)
12. Pedregosa, F., et al.: Scikit-learn: machine Learning in Python. J. Mach. Learn. Res. **12**, 2825–2830 (2011)
13. Radinsky, K., Horvitz, E.: Mining the web to predict future events. In: Proceedings of the sixth ACM International Conference on Web Search and Data Mining, pp. 255–264. ACM (2013)
14. Sayyadi, H., Hurst, M., Maykov, A.: Event detection and tracking in social streams. In: Third International AAAI Conference on Weblogs and Social Media (2009)
15. Zhang, T., Ramakrishnan, R., Livny, M.: BIRCH: an efficient data clustering method for very large databases. In: ACM Sigmod Record, vol. 25, pp. 103–114. ACM (1996)

From Univariate to Multivariate Time Series Anomaly Detection with Non-Local Information

Julien Audibert[1,2]([✉]), Sébastien Marti[2], Frédéric Guyard[3],
and Maria A. Zuluaga[1][iD]

[1] EURECOM, Sophia Antipolis, France
{audibert.julien,maria.zuluaga}@eurecom.fr
[2] Orange, Sophia Antipolis, France
sebastien.marti@orange.com
[3] Orange Labs, Sophia Antipolis, France
frederic.guyard@orange.com

Abstract. Deep neural networks (DNNs) are attractive alternatives to more traditional methods for time series anomaly detection thanks to their capacity to automatically learn discriminative features. Despite their demonstrated power, different works have suggested that introducing engineered features in the time series can further improve the performance. In this work, we present a feature engineering strategy to transform univariate time series into a multivariate one by introducing non-local information in the augmented data. In this way, we aim to address an intrinsic limitation of the features learned by DNNs, which is they rely on local information only. We study the performance of our combination compared to each individual method and show that our method achieves better performance without increasing computational time on a set of 250 univariate time series proposed by the University of California, Riverside at the 2021 KDDCup competition.

Keywords: Anomaly detection · Time series · Feature engineering · Non-local information

1 Introduction

A time series is a set of measured values that model and represent the behavior of a process over time. Time series are used in a wide range of fields such as healthcare [8], industrial control systems [2], and finance [15]. Detecting behavior or patterns that do not match the expected behavior of previously visualized data is a critical task and an active research discipline called time series anomaly detection [3,5]. Numerous methods to address this problem have been developed in recent years including statistical, machine learning and deep neural networks (DNNs) methods.

The performance of machine learning algorithms is correlated to the quality of the extracted features [14]. Feature engineering for augmenting time series

© Springer Nature Switzerland AG 2021
V. Lemaire et al. (Eds.): AALTD 2021, LNAI 13114, pp. 186–194, 2021.
https://doi.org/10.1007/978-3-030-91445-5_12

data is usually done by bringing external but correlated information as an extra variate to the time series. This, however, requires domain knowledge about the measured process. Another strategy is to create local features on the time series, such as moving averages or local maximum and minimum. Both strategies, as they are manual, are not very efficient, time consuming and require high domain knowledge expertise [7]. In theory, DNNs have emerged as a promising alternative given their demonstrated capacity to automatically learn local features, thus addressing the limitations of more conventional statistical and machine learning methods. Despite their demonstrated power to learn such local features, it has been shown that feature engineering can accelerate and improve the learning performance of DNNs [4].

In this work, we propose a novel feature engineering strategy to augment time series data in the context of anomaly detection using DNNs. Our goal is two-fold. First, we aim to transform univariate time series into multi-variate time series to improve DNNs performance. Second, we aim to use a feature engineering strategy that introduces non-local information into the time series, which DNNs are not able to learn. To achieve this, we propose to use a data structure called Matrix-Profile as a generic non-trivial feature. Matrix-Profile allows to extract non-local features corresponding to the similarity among the sub-sequences of a time series. The main contributions of this paper are:

- We propose an approach that transforms univariate time series into multivariate by using a feature engineering strategy that introduces non-local information to improve the performance of DNNs.
- We study and analyze the performance of this approach and of each method separately using the KDDCup 2021 dataset consisting of 250 univariate time series.

The rest of this paper is organized as follows. Section 2 briefly reviews other works on feature engineering for anomaly detection in time series. The Sect. 3 presents the transformation of univariate time series into multivariate one and the methods which constitute our framework. Section 4 describe the experiments and demonstrate the performance of our approach. The paper concludes with some discussion and perspectives in Sect. 5.

2   Related Works

Different studies have raised the importance of feature engineering for the detection of anomalies and the superiority of multivariate models in time series. A first study conducted by Carta *et al.* [4] shows that in network anomaly detection, the introduction of new features is essential to improve the performance of state-of-the-art solutions. Fesht *et al.* [7] compare the performance of manual and automatic feature engineering methods on drinking-water quality anomaly detection. The study concludes that automatic feature engineering methods obtain better performances in terms of F1-score. Ouyand *et al.* [11] shows that feature extraction is one of the essential keys for machine learning and proposes a method called

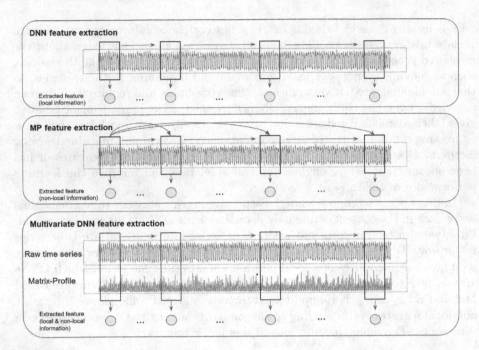

Fig. 1. Top: DNN automatic feature learning and extraction is limited to a local neighborhood, which is typically represented by the input window information. Middle: the matrix profile algorithm relies on non-local features, which are obtained by comparing every window of the time series. Bottom: the proposed strategy brings non-local feature information to a DNN by transforming the original univariate time series into a multivariate one by combining the raw time series and the non-local information obtained with matrix profile.

hierarchical time series feature extraction used for supervised binary classification. Finally, in [1], the authors conclude that multivariate models provided a more precise and accurate forecast with smaller confidence intervals and better measures of accuracy. Thus, studies have demonstrated the importance of feature engineering to improve anomaly detection models as well as the performance of multivariate methods compared to univariate ones on time series. Motivated by these ideas, our work aims to investigate how feature engineering using non-local information to achieve variate augmentation in time series can improve the performance of anomaly detection DNN models in univariate time series.

3   From Univariate to Multivariate Time Series

To take advantage of the performance of multivariate methods of anomaly detection on univariate time series it is necessary to transform the univariate time series into multivariate one. This can be achieved by adding external information to the time series, which requires specific domain knowledge. Our strategy, instead, transforms the univariate time series into a multivariate one, without any further information than the original time series, and is generic in that no specific knowledge on what the time series represents is required.

Our strategy consists in building another time series (i.e. another variate) by extracting non-local information from the raw time series, which DNN approaches fail to obtain as they typically operate in local neighborhood. To this end, we make use of the Matrix-Profile (MP) [16,17], a data structure for time series analysis. The proposed strategy is illustrated in Fig. 1.

The Matrix profile estimates the minimal distance between all sub-sequences of a time series. Thus, the Matrix-Profile value for a given sub-sequence is the minimum pairwise Euclidean distance to all other sub-sequences of the time series. A low value in the matrix profile indicates that this sub-sequence has at least one relatively similar sub-sequence located somewhere in the original series. In [9], it is shown that a high value indicates that the original series must have an abnormal sub-sequence. Therefore the matrix profile can be used as an anomaly score, with a high value indicating an anomaly.

In our approach, we propose to use the anomaly score obtained by Matrix-Profile over a given time series and merge it point-by-point with the original data. This can be thus seen as a data augmentation procedure using non-local information from the same signal.

As the new time series is just a multivariate time series, any given anomaly detection method can be used to identify anomalous points in it. In this work, we investigate three different estimation model-based techniques [3] as base anomaly detection methods. Among these category of methods, the auto-encoder [13] is among the most commonly used. An auto-encoder (AE) is an artificial neural network combining an encoder E and a decoder D. The encoder part takes the input window W and maps it into a set of latent variables Z, whereas the decoder maps the latent variables Z back into the input space as a reconstruction $\widehat{W}$. The difference between the original input vector W and the reconstruction $\widehat{W}$ is called the reconstruction error. Thus, the training objective aims to minimize this error. Auto-encoder-based anomaly detection uses the reconstruction error as the anomaly score. Time windows with a high score are considered to be anomalies [6].

Alongside the AE, we consider a more complex approach based on a Variational AutoEncoder (VAE) coupled with a recurrent neural network, the Long Short-Term Memory Variational Auto-Encoders (LSTM-VAE) [12]. In the LSTM-VAE, the feed forward network iof the VAE is replaced by a Long Short-Term Memory (LSTM), which allows to model the temporal dependencies. As in the AE, the input data is projected in a latent space. However, differently from the AE, this representation is then used to estimate an output distribution and not to simply reconstruct a sample. An anomaly is detected when the log-likelihood is below a threshold.

The third estimation model-based method we consider is denoted UnSupervised Anomaly Detection (USAD) [2]. USAD is composed of three elements: an encoder network and two decoder networks. The three elements are connected into an architecture composed of two auto-encoders sharing the same encoder network within a two-phase adversarial training framework. The adversarial training allows to overcome the intrinsic limitations of AEs by training a model capable of identifying when the input data does not contain an anomaly and thus perform a good reconstruction. At the same time, the AE architecture allows to gain stability during adversarial training of the two decoders.

The architecture is trained in two phases. First, the two AEs are trained to learn to reconstruct the normal input windows. Secondly, the two AEs are trained in an adversarial way, where the first one seeks to fool the second one, while this latter one aims to learn when the data is real (coming directly from the input) or reconstructed (coming from the other autoencoder). As with the base AE, the anomaly score is obtained as the difference between the input data and the data reconstructed by the concatenated autoencoders.

4   Experiments and Results

This section first describes the datasets used and the experimental setup used in our work. Then, we study the performance of our proposed approach and compare it against other techniques.

4.1   Datasets

In our experiments we use 250 univariate time series proposed by the University of California, Riverside at the 2021 KDDCup competition, consisting of univariate time series from many different fields. The 250 time series are composed of a training part containing data considered as normal and a test part containing one anomaly. The time series range from 6680 points for the smallest to 900000 points for the largest. The length of the training set represents on average 31% of the total length of the time series (i.e. a training on the first 31% points of the time series and a test on the next 69% points) with a minimum length of 2.5% and a maximum of 76.9%. All the time series are min-max normalized.

4.2   Experimental Setup

We use the percentage of correctly labeled series to assess the performance of our method. A time series is considered to be correctly predicted when the index of the point labeled as anomalous is included in a window of 100 points around the true anomaly.

We compare our method against the matrix-profile (MP), the auto-encoder (AE), the LSTM-VAE and USAD without the transformation of the time series. We compute the performance of the three anomaly detection methods AE, LSTM-VAE and USAD on a transformed univariate time series obtained using only non-local information, i.e. with Matrix-profile (MP-AE, MP-LSTM-VAE and MP-USAD). We assess both the AE, LSTM-VAE and USAD's performance using the proposed multivariate transformation, consisting of the original raw time series and the series obtained with MP, respectively (TS+MP)-AE,

Table 1. Hyper-parameter settings of the different methods

Method	Paramaters
MP	$window\_size = 100, discords = True$
AE	$window\_size = 100, latent\_dimension = 10, Epochs = 100$
LSTM-VAE	$window\_size = 100, Epochs = 100$
USAD	$window\_size = 100, latent\_dimension = 10, Epochs = 100$

Table 2. Methods performance and computational time.

Method	Performance	Train and Test time (s $\times 10^3$)
Matrix-Profile	0.416	1.47
AE	0.236	22.00
LSTM-VAE	0.198	85.31
USAD	0.276	29.00
MP-AE	0.292	22.16
MP-LSTM-VAE	0.344	84.30
MP-USAD	0.404	29.10
(TS+MA)-AE	0.148	22.38
(TS+MA)-LSTM-VAE	0.134	85.43
(TS+MA)-USAD	0.176	29.12
(TS+MP)-AE	**0.536**	22.50
(TS+MP)-LSTM-VAE	0.446	85.83
(TS+MP)-USAD	0.488	29.28

(TS+MP)-LSTM-VAE and (TS+MP)-USAD. To validate the relevance of the use of non-local information in the transformation of the time series, we also consider an identical combination with a local feature engineering strategy. In particular, in our experiments we use the moving average (MA), respectively (TS+MA)-AE, (TS+MA)-LSTM-VAE and (TS+MA)-USAD).

Implementation. We implement the AE using Pytorch and we used publicly available implementations for MP[1][1], LSTM-VAE[2] and USAD[3]. Table 1 details the hyper-parameter setup used for each method. Where a parameter is not specified, it indicated that we used those set by default in the original implementation

All experiments are performed on a machine equipped with an Intel(R) Xeon(R) CPU E5-2699 v4 @ 2.20 GHz and 270 GB RAM, in a docker container

---

[1] https://stumpy.readthedocs.io.
[2] https://github.com/TimyadNyda/Variational-Lstm-Autoencoder.
[3] https://github.com/robustml-eurecom/usad.

running CentOS 7 version 3.10.0 with access to an NVIDIA GeForce GTX 1080 Ti 11 GB GPU.

4.3 Results

Table 2 presents the results obtained by the different methods in terms of performance accuracy and computational times. Interestingly, we observe that the performance of DNN-based methods on univariate time series is very low and largely surpassed by the more conventional approach, the matrix profile. However, once the same techniques use the proposed data transformation strategy, we observe an important boost in their performance. The Auto-Encoder and the LSTM-VAE score almost 2.3 times higher when the combination of the matrix profile and real data is used as input instead of the original data. Similarly, USAD's performance increases by 1.8 times when the matrix profile and raw time series combination is used compared to its performance using only the raw time series.

Nevertheless, we observe that the non-local transformation alone is not enough to boost the performance of DNN methods. For instance, if the input consists only of the univariate time series transformed using the matrix profile, while there is some increased performance, this one is milder than when using a multivariate time series. This confirms that DNN methods perform better in a multivariate setup for anomaly detection.

Regarding the use of local features, i.e. the moving average, we observed that adding it does not allow USAD, LSTM-VAE and AE to increase their performance. Indeed, the combination of raw time series and moving average degrades the performance of AE and USAD by about 0.1 and the performance of LSTM-VAE by about 0.06. This suggests that any local features that might be discriminative can be extracted by the DNNs and introducing new manually crafted ones may be detrimental.

Finally, as it is expected, the computational time of DNN-based methods is much longer than the MP. However, what is interesting in our findings is that the computational time of DNN methods is very little impacted when the dimension of the time series increases. In fact, the AE's computational time goes from 21993 s in the fastest univariate configuration to 22491 s in the multivariate case. This means an increase of only 2.2% on computational time for a gain in performance of 230%.

5 Discussion and Conclusions

In this paper, we propose an approach to augment univariate time series using a feature engineering strategy that introduces non-local information in the generation of an additional variate to the series. In this way, we expect to address a limitation of DNNs, as they are not conceived to learn automatically non-local features. We achieve automatic non-local feature extraction by relying on the

Matrix-Profile, a method that computes the minimum pairwise Euclidean distance of all subsequences of the time series, and combining its output with the original time series.

We used data from the KDDcup 2021 competition containing 250 univariate time series to study the performance of our method. The performance analysis highlighted the relevance of transforming the univariate time series using the proposed feature engineering and data augmentation strategy. Our results show that introducing non-local information to augment the dimension of the series improves the performance of DNN methods. For instance, by using a very simple method, such as an autoencoder, we were able to obtain a gain in performance of 230%, without significantly increasing the computational time. As such, our preliminary results suggest that non-local information represents an important source of additional information that can increase performance of DNN methods.

While our approach focuses on the particular case of transforming uni- to multivariate time series, this idea could be used to augment time series, which are multivariate at origin, as a way to introduce non-local information.

In this work, we used three methods of anomaly detection based on Deep Neural Networks in combination with Matrix profile. The good performance on a simple auto-encoder, a recurrent network such as LTSM-VAE and USAD, a state-of-the-art neural network, suggest that our combination could generalize to other DNN methods. Therefore, future works should explore other feature engineering techniques that can provide non-local information, as well as other multivariate DNN anomaly detection methods.

Finally, our findings are consistent with one of the results of the time series prediction competition, the M4 challenge [10], which highlighted the predictive power of ensemble approaches combining learning-based with more conventional statistical methods. Due to the great success of DNN methods in the recent years, it is now often the case that more traditional methods are overseen. Our results suggest that the use of hybrid approaches should be further explored.

References

1. Aboagye-Sarfo, P., Mai, Q., Sanfilippo, F.M., Preen, D.B., Stewart, L.M., Fatovich, D.M.: A comparison of multivariate and univariate time series approaches to modelling and forecasting emergency department demand in western australia. J. Biomed. Inform. **57**, 62–73 (2015)
2. Audibert, J., Michiardi, P., Guyard, F., Marti, S., Zuluaga, M.A.: USAD: unsupervised anomaly detection on multivariate time series. In: Proceedings of the 26th ACM SIGKDD International Conference on Knowledge Discovery and Data Mining, KDD 2020, pp. 3395–3404. Association for Computing Machinery, New York (2020)
3. Blázquez-García, A., Conde, A., Mori, U., Lozano, J.A.: A review on outlier/anomaly detection in time series data. ACM Comput. Surv. (CSUR) **54**(3), 1–33 (2021)
4. Carta, S., Podda, A.S., Reforgiato Recupero, D.R., Saia, R.: A local feature engineering strategy to improve network anomaly detection. Future Internet **12**(10), 177 (2020)

5. Domingues, R., Filippone, M., Michiardi, P., Zouaoui, J.: A comparative evaluation of outlier detection algorithms: experiments and analyses. Pattern Recogn. **74**, 406–421 (2018)
6. Fan, C., Xiao, F., Zhao, Y., Wang, J.: Analytical investigation of autoencoder-based methods for unsupervised anomaly detection in building energy data. Appl. Energy **211**, 1123–1135 (2018)
7. Fehst, V., La, H.C., Nghiem, T.D., Mayer, B.E., Englert, P., Fiebig, K.H.: Automatic vs. manual feature engineering for anomaly detection of drinking-water quality. In: Proceedings of the Genetic and Evolutionary Computation Conference Companion, GECCO 2018, pp. 5–6. Association for Computing Machinery, New York (2018)
8. Kale, D.C., et al.: An examination of multivariate time series hashing with applications to health care. In: 2014 IEEE International Conference on Data Mining, pp. 260–269 (2014)
9. Linardi, M., Zhu, Y., Palpanas, T., Keogh, E.J.: Matrix profile goes mad: variable-length motif and discord discovery in data series. Data Min. Knowl. Disc. **34**, 1022–1071 (2020)
10. Makridakis, S., Spiliotis, E., Assimakopoulos, V.: The M4 competition: 100,000 time series and 61 forecasting methods. Int. J. Forecast. **36**(1), 54–74 (2020)
11. Ouyang, Z., Sun, X., Yue, D.: Hierarchical time series feature extraction for power consumption anomaly detection. In: Li, K., Xue, Y., Cui, S., Niu, Q., Yang, Z., Luk, P. (eds.) LSMS/ICSEE -2017. CCIS, vol. 763, pp. 267–275. Springer, Singapore (2017). https://doi.org/10.1007/978-981-10-6364-0_27
12. Park, D., Hoshi, Y., Kemp, C.C.: A multimodal anomaly detector for robot-assisted feeding using an LSTM-based variational autoencoder. IEEE Robot. Autom. Lett. **3**(3), 1544–1551 (2018)
13. Rumelhart, D.E., Hinton, G.E., Williams, R.J.: Learning Internal Representations by Error Propagation, pp. 318–362. MIT Press, Cambridge (1986)
14. Soni, A.N.: Feature extraction methods for time series functions using machine learning. Int. J. Innov. Res. Sci. Eng. Technol. **7**(8), 8661–8665 (2018)
15. Theodossiou, P.T.: Predicting shifts in the mean of a multivariate time series process: an application in predicting business failures. J. Am. Stat. Assoc. **88**(422), 441–449 (1993)
16. Yeh, C.M., et al.: Matrix profile I: all pairs similarity joins for time series: a unifying view that includes motifs, discords and shapelets. In: 2016 IEEE 16th International Conference on Data Mining (ICDM), pp. 1317–1322 (2016)
17. Yeh, C.C.M., Kavantzas, N., Keogh, E.: Matrix profile VI: meaningful multidimensional motif discovery. In: 2017 IEEE International Conference on Data Mining (ICDM), pp. 565–574. IEEE (2017)

Author Index

Printed in the United States
by Baker & Taylor Publisher Services